MUSIC BETWEEN YOUR EARS

music between your ears

HOW MUSICAL ENGAGEMENT POWERS THE HUMAN BRAIN

Samuel Markind, MD

JOHNS HOPKINS UNIVERSITY PRESS
Baltimore

Graphic design work by Lou Okell, Arkettype Graphics (www.arkett.com).

Johns Hopkins University Press
2715 North Charles Street
Baltimore, Maryland 21218
www.press.jhu.edu

Library of Congress Cataloging-in-Publication Data is available.
A catalog record for this book is available from the British Library.

ISBN 978-1-4214-5237-1 (hardcover)
ISBN 978-1-4214-5238-8 (ebook)

*Special discounts are available for bulk purchases of this book. For more
information, please contact Special Sales at specialsales@jh.edu.*

EU GPSR Authorized Representative
ÐOGOS EUROPE, 9 rue Nicolas Poussin
17000, La Rochelle, France
e-mail: Contact@logoseurope.eu

For Dina
The love of my life
The focus of my world

Many women have done excellently,
but you surpass them all.
—Proverbs 31:29

The woman by my side is the fruit
of the Garden of Eden.
—David Broza

Contents

Foreword

Sheilah Rae

When an email from Dr. Sam Markind arrived in my inbox, I was surprised. I usually hear from him only when there's an upcoming book for our book group. I don't know him professionally, and while he and my husband, Elliott, are in the same medical field, they aren't close colleagues. So, when Sam asked if I might be interested in writing the foreword for his new book on music and the brain, I must tell you, I was stunned. Honored, but stunned.

I've been a professional musician and performer for over seventy years, having made my professional debut as a singer and dancer at the age of ten. Also, I've been married to a neurologist for more than fifty years. To Sam, I guess that was reason enough to invite me to do this, but then he reminded me of something from our past. His request had brought us both full circle! (More on this in a bit.) At that point, I figured, if this is good enough for Sam, it's good enough for me. And so, dear reader, here I am.

I come from a big musical family: My grandfather was a cantor. My father's oldest sister was a concert pianist who made her debut with the Chicago Symphony at sixteen. My mother's father as well as her stepfather were violinists. My mother herself was a fine pianist and singer but gave up both when my three sisters and I came along. Among my sisters, two became professional dancers and choreographers; the third is now a retired dance history professor.

My daughter, a composer and singer in Los Angeles, grew up playing with her Barbies under our Steinway while I was banging away on some new song or jingle assignment. Years later, she told me that until she was ten, she thought all adults played the piano since everyone coming to our home seemed to do so. Our son is extremely musical as well but decided to pursue other dreams. Still, one of my favorite photos is of him playing "Twinkle, Twinkle, Little Star" on his Suzuki violin at his fifth birthday party.

My parents believed that if you were going to do something—anything—you went to the best place to study it, and I became a beneficiary of this generous philosophy. I trained at the Chicago Musical College through high school, then the University of Michigan School of Music. After that, I attended the Royal Academy of Music and the Guildhall School of Music and Drama in London. I studied voice, piano, guitar, violin, and dance—lots and lots of dance.

After years as a Broadway performer and with a young family of my own, I segued into songwriting. It was a logical evolution. It meant I could spend more time with my kids and experience the joy of hearing others sing my songs. To this day, both are still a thrill.

What I know about the technical aspects of music in the brain is minimal, limited to dinner parties discussions over the decades with our musician friends plying Elliott with questions about memory—how we learn, where melodies are stored, and the like—some heady stuff over Sancerre and grilled salmon. At this point, I could probably do a mean neurological exam, after years of watching Elliott as I helped in the office, or what we musicians call "the survival job." As for deciphering the findings of an exam, however? That's the magic in neurology. Only those trained and experienced can intuit what they mean.

To achieve excellence in medicine, as in music, requires years of practice and experience. In that regard, Elliott and I are alike—disciplined in our fields and continually honing our abilities and understanding. One of Elliott's oldest friends was the late great Oliver Sacks, who not only read the Oxford English Dictionary for fun—I'm not kidding, and he could quote from it too—but could then turn around and play a perfect Debussy etude with the sensitivity and nuance that would have astounded many of my musician colleagues. I mention him here because he is the only other neurologist I have known personally who wrote about music and the brain and was involved in early clinical trials for music therapy. His many books cover some of the ideas about how we learn, and what music means to us, particularly in patients with Parkinson's disease or dementia. His theses did not, however, deal with the kind of specificity in Dr. Markind's premise: Music is in all of us; it is a uniquely evolutionary part of our DNA and contributes to our survival as humans. I love this idea. Although most of us aren't going to become professional musicians, it's comforting to know that we all have this link to one another, that on some level our brains are (forgive the pun) harmoniously attracted to each other because of it.

So why read this book? Are you going to understand it? The simple answer is yes. Dr. Markind has written a scientific, but totally accessible study of how the brain processes music. His idea for it came from two lectures he gave on the subject, in Santa Fe and Paris. Both were remarkably well received, so much so that it encouraged him to do a deeper dive into how music works between our ears. A couple of years ago, Elliott and I attended an early "rehearsal," where Sam shared the nucleus of some of this material with a few friends. This is what I meant by my having come full circle: I was there when Sam was beginning to explore this subject, and now, here I am sharing the fruits of his exploration with you.

At the rehearsal, Sam also bravely sat at the piano and demonstrated some of his musical thoughts, while telling us he needed to practice more before his talk in Santa Fe. We all laughed with him at his honesty, which brings me to one of the other reasons for this book—Sam's marvelous, all-encompassing call to action: Engage in music or dance in your life. You don't need to be a professional musician or dancer or anything close to that. Put on a CD and have a dance-a-thon with your kids. Sing in the car or the shower. Teach your kids how to sing, even if it's just a simple "round." I can't tell you the number of hours I spent with our children when they were young, singing "Row, Row, Row Your Boat," with all of us giggling when we lost our place. You will be richer for it. Your life will be fuller. You'll be happier. And it will keep you young—young at heart, young in your mind, and young physically.

So . . . Anyone care to join me in a little jitterbug?

Sheilah Rae is a gifted lyricist, composer, and librettist. In recognition of her career of accomplishments, the New York Theatre Barn honored her with their Lifetime Achievement Award. For more information, visit her website, www.sheilahrae.net.

Preface

When I reflect on my journey to date through the cycle of life, I appreciate that I have had many fine teachers. Three in particular stand out, and they encouraged me in ways that bear upon my writing this book. I wish to share them with you, the readers of this book. Great and engaging educators in their individual ways, all go beyond mere instruction. They communicate that they care about the people they educate, and they have natural curiosity, making them open to exploring new ideas and meeting new people. In telling you about them, you will gain an understanding of how this book came into being.

Dr. Barry Panter ignited a spark in me through the unique Creativity and Madness conferences he founded in 1981. Every year, hundreds of people attend the conference, usually held in Santa Fe, drawn to its novel presentations and workshops focused on the junction of health care and the arts.

An opera-loving physician, Barry related the serendipitous origin of the conference: One day he returned to his and his wife's hotel room in Hawai'i and complained to her about being bored while attending a traditional, continuing medical education seminar.

"Why doesn't someone put on a conference that's interesting, that the spouse can go to, that they can sit . . . at dinner and talk about what they've heard and learned?"

His wife simply responded, "Why don't you do it?"

Creativity and Madness conferences were the offspring of that challenge.[1]

Barry also publishes books, collections of some of the best conference presentations. I stumbled across an advertisement for one of these books in 2005 and, after purchasing it, found myself on Barry's mailing list. Invitations to attend the annual conference began to arrive in my mailbox. After ignoring them for a couple of years, I said to my wife after receiving the invitation in 2008, "Let's go check this conference out." So, off we went to Santa Fe. The trip opened new worlds to us.

To begin with, New Mexico is relatively unknown country to most people living in the northeastern United States. Santa Fe is a highly engaging town, culturally as well as historically. The nearby towns of Taos and Los Alamos boast artistic and scientific import, respectively. For the moment, Albuquerque remains underrated, especially its delicious cuisine.

Creativity and Madness attracts a large contingent of mental health care professionals, so there I was, a neurologist, swimming in a sea of psychiatrists, psychologists, psychiatric social workers, and the like. I was surprised that many of them operated from a psychoanalytic perspective. I attended a small Q&A session led by Barry and one of the other presenters. Questions about the id and mothers were being discussed. After a while I felt compelled to raise my hand. "What do you think about a neurobiological basis of creativity?" I asked. What happened in the next moment reoriented my professional mindset. Barry looked me straight in the eye and said, "Sam, you're a neurologist, what do you think?" It caught me totally off guard. I had no idea Barry knew who I was, given that I was attending for the first time and was one among hundreds of other attendees. I didn't know how to respond. After reflecting on the question for a few days, I became determined to answer Barry's challenge by returning to Santa Fe to present a session.

Four years later, I gave a presentation on creativity from a neurobiological perspective. To prepare, I had read voraciously and educated myself a great deal on the subject. I also found

Barry a pleasure to work with. His goal is to stimulate people passionate about the crossroads of health care and the arts to research the heck out of their chosen subject and then give a lively and engaging presentation about it. Beyond those basic guidelines, everything else rests in large measure on the presenter's imagination. Each presentation I've done has increased my appetite to do another.

Given the overlap of health care with the arts at the Creativity and Madness conferences, I decided to weave live piano music into my scientific talks. To accomplish this, I needed help. Fortunately, Paul Nickolloff stood by my side every step of the way. I met Paul when I was looking for a piano teacher. With a background in general education as well as in music education, his skill set encouraged fabulous dialogue between us.

After attending the 2008 conference as a spectator, I decided to create a neurologically oriented presentation on creativity—a natural choice for me, since I'm a neurologist—and incorporate live music into it, Gershwin tunes to be specific. George Gershwin is a fantastic subject for a presentation. He not only composed fabulous songs, but also has an interesting neurological history. And there are tons of stories about him—perhaps half of them are true, but no one knows which half.

Thanks to a large dose of Paul's help, I created a musical component to complement my medical lecture. With his instruction, I learned to accompany myself, how to string songs together to construct a medley, and how to create a successful sing-along. Meanwhile, I diligently researched the neurology of creativity. I also practiced, practiced, and practiced my talk. The presentation proved to be a wonderful success and made me thirsty to do it again. First, however, I needed a rest. The research, the imagination, and the practicing required to assemble such a program—all while working full-time and raising a family—was exhausting. It took me nearly a year to rest and re-charge. Then I was back at it, inspired by the idea of combining Aaron Copland's musical biography with a study of his dementia, in particular exploring when his composing first suggested a decline in creativity. My concept was that a change in creativity—typically a decline—is an early sign of a dementing illness. Paul and I studied Copland's works and established a timeline to demonstrate when a decline in creativity began to show itself in his compositions.

Meanwhile, with Paul's help, I dove into playing Copland's music, putting together several numbers to play during my presentation, and gained a profound appreciation for its sensitivity and originality. During this process, I became acquainted with the Copland House concert series directed by the renowned pianist Michael Boriskin, and as a result, connected with a member of the Copland family. Through this contact, I got the opportunity to visit the music rehabilitation center that Oliver Sacks had established, and I met Dr. Concetta Tomaino, an internationally respected music therapist.[2] Talk about good fortune! Sadly, Dr. Sacks died about a year before I visited. After another successful presentation in Santa Fe, it was time for another break while awaiting inspiration.

One day, probably while I was taking a shower, the idea came into my head to create a presentation devoted to examining the impact of music on the brain. Many books have been written on this general theme, but I felt that a sharp focus on the subject from an evolutionary perspective had yet to be fully explored. It was also a fantastic topic for collaborating with Paul, given his insights into both education in general as well as music education; at that time, he was the principal of a middle school. With Paul's input, I assembled my presentation, collecting such a wealth and breadth of information that I came to realize it could also be the seed of a book. This volume is the fruit of this realization.

Desiring to work live music into the program that I gave in 2021, I searched for a musical theme with a twist, something out of the ordinary. I landed on the idea of incorporating French chanson–era music that I like; surely, I figured, this would be different from anything previously performed at Creativity and Madness. I focused on Charles Aznavour, the greatest male vocalist and songwriter of the era. In so doing, I discovered the incredible breadth of Paul's musical knowledge. He was already familiar with many of Aznavour's songs as we worked on integrating several of them into the program.

This brings me to a posthumous tribute to Linda Riblet, my high school French teacher for three years. She was the most outstanding and memorable teacher I had in high school. Madame Riblet had a number of funny sayings, the most unforgettable being "Translations are like women: If they're beautiful they're not faithful, and if they're faithful they're not beautiful." I suspect this saying wouldn't go over too well nowadays, but in the mid-1970s, here was an adult woman (and mother!) saying this to a class of adolescents. My mom laughed when I came home from school that day and told her about it.

Madame Riblet had the ability to keep adolescents on track, no easy feat, and I diligently studied what she taught. While her French language lessons made a mark on me, her life lessons were even more powerful. She explained that the main reasons to learn a foreign language are to expand our world, through learning about ideas we would never be exposed to otherwise, and to connect with people we would otherwise never know. She even went so far as to suggest that one day we would meet a Francophone who didn't speak English and would need us to translate. I thought her idea was way over the top when she said this, yet four years later, that's exactly what happened.

I found myself on a bus tour to the region of the Dead Sea, and en route the guide asked if anyone could translate English to French for someone on the tour. No one else's hand went up, so the task—heck no, the opportunity—fell into my lap. I spent the day with Sophie, trying my best, with my limited French vocabulary, to explain what the guide was saying. By the end of the day, my exhausted mind was in a mist, but I remembered to ask Sophie for her address. We wrote letters—this was well before email—and, two years later, I visited her family in Paris. Ever since, I have remained in contact with her parents and her brother, along with his wife and their children after they entered the picture.

Meeting this lovely family kindled a keen desire in me to study French in a more serious way so I could have meaningful conversations with French speakers wherever I might meet them. It has also given me a front row seat to the movable feast called Paris and marked the beginning of a voyage because knowing a language opens the door to a culture. In particular, that was my point of departure into Aznavour's music, especially his wonderful song "La Bohème" about the unique historical time and place known as bohemian Montmartre.

To this day, I often think back to what Madame Riblet remarkably predicted those many years ago. One of my biggest regrets is never contacting her to tell her about meeting Sophie & family. I'm certain she would have been thrilled and felt great joy in hearing about this adventure, the kind of appreciation that makes a teacher feel truly fulfilled. Years passed, and the thought briefly popped into my mind from time to time, but I never reached out. Finally, after decades had gone by, I tried to locate her only to discover that she had passed away a few years earlier. Dear reader, if the thought to thank one of your teachers ever crosses your mind, please don't hesitate. Reach out and do it. You will be glad you did, and I'm sure your teacher will be delighted.

Right after my 2021 Creativity and Madness presentation, one of the conference attendees approached me. She identified herself as an organizer of a scientific conference to be held in Paris the following year and invited me to give my presentation there. How could I say no? I invited my Parisian family to the lecture. Five of them came, including a member of the "younger" generation who had recently started studying neuroscience at university. It was as if everything in its way is connected to everything else in the cycle of life.

MUSIC BETWEEN YOUR EARS

Introduction

Welcome to an exciting journey to the junction of music and the brain. A voyage through either the brain or music is terrifically interesting in its own right, and looking at the relationships between the two makes it even more so.

My primary goal in writing this book is to convey music's powerful effects on the human brain from an evolutionary perspective. In a nutshell, music is valuable to the brain by virtue of its ability to help humans survive and procreate, the paramount aims of evolution. My second goal is to encourage people who do not actively engage with music to do so. Music's impact is far greater on a person who joins in, and we'll explore ways to lower barriers to participation.

Members of the scientific community span the gamut regarding the importance they attach to music's place in the lives of humans. At one end of the spectrum are those who deny the notion of music having any worth, dismissing it as an accidental byproduct of language, or as one scholar refers to it, "auditory cheesecake."[1] At the other end of the spectrum are those who deem music to be one of the principal propulsive forces of human development, who believe "it makes us human."[2] Wherever one stands along this continuum, it is clear that music has the power to generate passionate discussion. Indeed, members of the artistic community may well say that evoking passions is music's highest calling.

THE BRAIN SCIENCE OF MUSIC TAKES SHAPE

Modern comprehension of the brain's anatomy began to emerge around the middle of the seventeenth century. Medical science's understanding of the relationship between music and the brain originated two centuries later, when contemporary concepts about how the brain works started to take hold. Progress was steady, but slow, as advancing knowledge relied heavily on autopsies. The playing field changed dramatically in the 1970s, when imaging of the living brain arrived on the scene, beginning with computed tomography (CT) scanning, which was largely developed at a British record company, EMI.[3] Magnetic resonance (MR) scans began to come into general use in the 1980s, providing another method to picture the brain.

Initially, scanners were used to image abnormal anatomy and their utility was limited to patient care. Subsequent technological improvements enabled the visualization of physiology in real time, making it possible to study the functioning of the brain in both health and sickness.. These advances empowered scientists to observe how the brain processes various components of music and opened the door to examining fundamental questions about the relationship between music and the brain. A whole new discipline—the brain science of music—came into being as a result. Numerous conferences were organized, bringing the growing community of scientists studying music and the brain together. The proceedings of one such conference, dedicated to promoting scientific inquiry into the interrelationship of music and the brain, led to *The Cognitive Neuroscience of Music* (2003),[4] edited by Isabelle Peretz and Robert Zatorre, a landmark volume of specialized articles written by and for a scientific audience.

Books geared to the general public also appeared, several of which enjoyed wide appeal. Two merit specific mention, the first being *This Is Your Brain on Music* (2006). Its author, Daniel Levitin, worked in the music business before becoming a scientist studying music and the brain. Highly informative with respect to how the brain perceives and produces music, it is also entertaining, filled with anecdotes about people in the music business, many of them household names. The second book, *Musicophilia* (2007), by Oliver Sacks, stands as an extraordinary achievement by arguably the greatest author in the field of neurology. It can serve as a reference text on the brain and music for health care professionals, yet Sacks' breezy case study method is so human and approachable that a general audience can take pleasure in reading it.

I read these books, and many others as well, and enjoyed and learned something from all of them, but each one left me feeling that it had failed to squarely address the issue gnawing at my mind: Does music provide evolutionary benefit? That is, does music promote survival and procreation of the human species, or in the terminology of evolution, does music further natural selection and sexual selection? As noted above, probing this question is my primary goal in writing this book.

The theory of evolution posits survival and procreation as life's paramount goals. All living species face the challenge of a perpetually changing environment, and to survive, species must either adapt to the changes or find a new home in a more hospitable environment. The principal currency of adaptation is genes, traits coded by DNA (deoxyribonucleic acid) in the nucleus of our cells. Traits that enhance the ability of a species to endure and reproduce have high "survival value,"[5] ensuring that they will be consistently passed down from generation to generation. On the other hand, traits that diminish the likelihood of survival and procreation will be eliminated; sometimes entire species become extinct on account of this. Traits that neither promote nor reduce the likelihood of survival and procreation may be inherited in a random fashion.

Over eons, learning arose as a method for transmitting new adaptive skills far more quickly than the time it takes for a genetic mutation to appear. Species could therefore adjust to changing circumstances or novel environments with greater success. Unlike genes, however, learning cannot be easily passed on to the next generation. Learning demands a program of dedicated teaching and repetitive practice to be transmitted to others. Humans have excelled at learning far beyond other species thanks to two extraordinary ways to transmit adaptive skills: music and language, which together form core elements of culture. That music is found in every human society speaks to its importance in this role.

Beyond the tremendous knowledge brain music science provides, it also teaches that actively engaging with music offers far greater advantage than passively consuming it. This is the power of sing, dance, and play to effect positive change in one's life. For people who do not currently participate with music actively, I aim to encourage them to do so. I explore music participation through the themes of sing, dance, and play, with the goal that one or more of these will appeal to you, the reader, and lead you to active engagement with it. My suggestions are straightforward and practical. I'm not a music professional, but I enjoy music actively at levels appropriate to me and want others to find ways to enjoy it actively at levels appropriate to them. I will consider this book to be a success if it increases the number of people who actively engage with music.

NOTE TO READERS

This book addresses several different audiences. The first is the array of health care providers interested in the intersection of music and the brain. This is the broad group to which I belong and also the target group for presentations I gave that became the genesis of this book. I believe I speak for many of us in the field when I say that health care has become increasingly based upon computer language algorithms to address patients' problems. One result of this is a devaluation of music in the healing process as well as for health maintenance. My hope is that the information, stories, and suggestions relayed in this book lead providers to increase their engagement with music in their work and spur them to promote a greater degree of participation in music to patients and clients as part of their health maintenance or healing process.

The second, closely related audience consists of music researchers. Members of this field have done a marvelous job over the roughly three preceding decades of unveiling the neuroscience of music, identifying how the brain and music interact with one another. I believe the time has now come to narrow the gap between music research and clinical practice. As Professor Hervé Platel, a renowned music neuroscientist, points out in chapter 1, music has a therapeutic potential to enhance underperforming brain functions since it so powerfully stimulates the brain. This opens the door to music researchers and clinicians to join hands to improve health outcomes. It could benefit people currently in good health as well as patients rehabilitating from illness or injury.

Music educators constitute the third audience for this book. Although not qualified to advise on the "how" of music curricula, I hope to provide some of the "why" for supporting music curricula. I've had many conversations with my friend Paul Nickolloff about music education. With a background in both general and music education, he has raised my awareness of the persistent decline of music education in schools as well as in society at large. This awareness led me to invite Scott Shuler, a past president of the National Association for Music Education, to write the Afterword to this book. Experienced in curriculum development at the state and national levels, he cochaired the development of the United States' current National Standards in Music Education. I look forward to this book contributing in some measure to supporting music education.

For people in the music business, the fourth group, I recognize that their conversations with brain scientists can be daunting, even when both parties are interested in the subject. Oftentimes, they are simply speaking different languages. Given this, I asked Sheilah Rae, musically accomplished both in vocals and in composition, to write the book's Foreword. Being married to a neurologist as well as having received the 2019 Lifetime Achievement Award from the New York Theatre Barn for her exceptional contributions to musical theater, she offers a unique combination of success in the music business and longstanding exposure to brain science.

The fifth audience is the broadly informed general reader. I view you as a person who reads for the love of learning something new and interesting. There is hardly anything more interesting to learn about than the brain and more engaging to read about than music. My hope is that you take away something of value about music, the brain, or both from this book and then share it with your friends and family.

ORGANIZATION OF THE BOOK

This book is organized into three parts. The first, A Musical Brain, begins with an exploration of the fundamentals of music and the brain in chapter 1, avoiding technical jargon as much as possible. Chapter 2 takes a deeper dive into the relationship between music and the brain, drawing on more technical terms along with helpful illustrations. Those who hesitate to swim in such waters may defer chapter 2 and return to it later, after reading the second and third parts of the book. The first two chapters demonstrate that the brain works in fascinating ways often hidden from human awareness and that music is ultimately what the brain makes of it.

Chapters 3, 4, and 5 form the second part of the book: The Evolutionary Impact of Music on the Brain. This section examines the evolutionary effect of music on the brain from three different angles. Chapter 3 discusses the brain's innate capacities, abilities with which we are born, for music. It begins with a look at studies of babies that reveal their brains to be naturally "wired" for music even before they enter this world. Later in life, however, people may lose aspects of music function as a result of illness or injury. Such disorders also reveal important insights into how the brain processes music.

The intriguing relationship of music and language is then examined. On the one hand, certain brain functions serve both of them, but on the other hand, differences between music and language require the application of distinct skills from different brain regions, some of them at a distance from each other. I believe music and language coevolved. The underlying brain capacities specific for each arose in response to the physics of sound. As the human species evolved, the music capacity of the brain and its language capacity advanced alongside one another. Each developed to address different human needs.

Chapter 4 explores needs for which music developed, presenting a variety of ways—I call them hypotheses—that music confers evolutionary benefit on humans, increasing the likelihood of survival and procreation. The first three ways relate to natural selection: communication, while typically thought of as the hallmark of language, music can be the optimal way to communicate information in certain contexts; caregiving, including ways in which music facilitates raising children, the confluence of survival and procreation; and entrainment, the coordination of bodily movement with music, a fascinating attribute that enhances productivity and socialization. Shifting orientation, the focus then turns to music's intimate association with sexual selection, which Darwin considered to be the prime driver of music's development.

Chapter 5 looks at how music can improve quality of life. This applies to normal brains as well as to brains suffering from illness. This chapter emphasizes the role music can play in learning, that is, educating and reeducating brains. Musical interventions to improve three specific neurologic conditions—disorders of language, movement, and memory—are also discussed. This chapter features several personal stories about how music has helped people recover from a variety of physical and mental health challenges.

No matter how much neuroscience is surveyed, however, music will always remain a personal experience and sometimes profoundly so. At the same time, it will also forever be of the brain and of the heart.[6] It has the power to lift us up, to feel great joy, or to help us persevere through profound sorrow. It can touch the very core of one's personal identity, bringing each person in touch with their own story—including my story, which is important and real to me, as well as your story, which is as important and real to you.

Moreover, no matter the circumstances, music's power is greatest when you actively engage with it. The third part of this book, chapter 6, is a coda, a musical ending, geared to readers who do not yet participate actively with music. In this epilogue, I offer ideas for bringing music into everyday life, grouped into the three categories noted above: sing, dance, and play. One commits a personal disservice by thinking that "doing" music is only the province of talented, elite performers. There was a time, not so long ago, when almost everyone engaged actively with music in one way or another. Bringing that back as a facet of music would benefit many people physically, socially, and emotionally. My hope is that lowering the barriers to participation can help make active engagement with music more accessible and attainable.

Appendix A presents a diagram of the music network described by the outstanding neuroscientists at the University of Helsinki in Finland. The network contains the structures of the brain most essential for music function. Appendix B is a compilation of renowned neurologic physicians and music scientists mentioned throughout the book.

Even when physicians and scientists avoid technical terms as much as possible, their manner of speaking can be challenging to the lay reader. Therefore, a glossary is included to help with technical, musical, and medical terms. Each term in the glossary is bolded on its first appearance in the text. Finally, in order to enhance readability, many of the neuroscientific details and historical references have been placed in the endnotes. Readers interested in expanding their knowledge of these fascinating aspects of the discussion are invited to refer to the endnotes frequently.

Now it's show time! Let's delve into the power of music on the brain.

PART I

A Musical Brain

CHAPTER 1

The Basics of a Musical Brain

"Without music, life would be a mistake."
—*Friedrich Nietzsche*

Music is ancient, but no direct evidence points to when humans first began to make it. Anthropologists who study fossil data continue to debate when humans developed the anatomy necessary to sing (throat structure and breath control) or dance (erect stance on two legs and balance function). Some claim it occurred as far back as 350,000 years ago, but others argue not until around 40,000 years ago.[1] Even if the date were known when humans first developed the physical apparatus to make music, it would provide no proof that they sang or danced or played an instrument immediately upon gaining the ability to do so.

Three "flutes," unearthed as a set in the Geissenklösterle Cave, in southwestern Germany, are the oldest artifacts clearly identified as musical instruments.[2] They have been dated between 40,000 and 45,000 years ago.[3] One of the instruments, played by blowing into a hollow tube, consists of mammoth ivory, the other two of swan bone. The Hohle Fels flute, made of vulture bone and the most complete of all the ancient bone flutes found so far, also comes from southwestern Germany and is dated between 35,000 and 40,000 years ago. About 8.5 inches long, it closely resembles a modern-day flute and, because the part of the instrument into which the musician blows remains intact, ranks as an especially valuable discovery.[4]

The oldest known drum, found in China, dates to around 5,000 or 6,000 years ago. People likely made sounds by banging on rocks, called lithophones when used in this way, for many thousands of years before then. Striking an animal skin stretched across a wooden frame, a membranophone, appeared later than flutes made of animal bones.[5]

Why did ancient people make these instruments? Were they the tools of gifted musicians "reaching for the stars," playing moving, soulful melodies and rhythms similar to a modern tune produced with a recorder and drum? Did a clan discover that music helped it to "keep its feet on the ground" by coordinating the movements of combat or hunting parties? After all, armies marched into battle to the music of fifes and drums until just two centuries ago, and paleolithic cave paintings of large game animals are preferentially located in the most vocally resonant locations of caves.[6] Alternatively, some researchers believe that building community is music's greatest contribution to humanity and that ancient musical instruments helped early humans develop and maintain strong interpersonal bonds.

Ancient literature is full of references to music and the role it plays in human society. Music has often been viewed as a particularly effective means to connect with the Divine. Here are a few examples from the Western tradition, beginning with a selection from Psalms, collections of verses designed to be sung.

Psalm 98
Sing to the Lord a new song for He has worked wonders. . . .
    Break into joyous songs of praise. . . . Sing praise to the Lord
    with the lyre and melodious song.[7]

Music played a central role in the social and cultural life of ancient Greece. Mixed-gender choruses performed for entertainment, celebration, and spiritual (religious) ceremonies. Music was one of the major subjects taught to children; in fact, the Greeks considered musical education important to the proper development of an individual's mind and soul.[8]

Homeric Hymn to Apollo
They walked, and Lord Apollo, son of Zeus, led them, playing
a lovely song with lyre in hand, stepping fine and high.[9]

From Homer's *Iliad*
All day the youths of the Achaeans long propitiated the god
with song and dance, singing a beautiful paean [that]
delighted [the god's] heart when he heard it.[10]

Centuries later, early Christian leaders also found music to be valuable.

Ephesians 5:19
[Address] one another in psalms and hymns and spiritual
songs, singing and making melody to the Lord with all your
heart.[11]

Early civilization in the Indus Valley had a parallel tradition. An ideogram found there contains the earliest known depiction of an arched harp—an instrument played by plucking one or more strings—dated to nearly 4,000 years ago.[12] Statuettes from this epoch depict male and female figures dancing. Sanskrit scholar and musicologist V. Ragahavan states, "Our [Indian] music tradition in the North as well as in the South remembers and cherishes its origin in the Samaveda . . . the musical version of the Rigveda [sacred Hindu foundational document]."[13]

Egyptian tomb and temple paintings depict numerous illustrations of musical instruments being played. A fine example is the Musicians of Amun from the tomb of Nakht (Eighteenth Dynasty, Thebes).[14] This ensemble of three women includes two musicians plucking on string instruments resembling a modern ukulele and harp, while the third plays a wind instrument suggestive of a recorder.[15] At more than 3,000 years old, the pair of silver trumpets discovered in Pharaoh Tutankhamun's tomb by Howard Carter in 1922 are the only trumpets known to have survived from ancient Egypt.[16] They are believed to be the oldest playable trumpets in the world.[17]

As shown here, it is well documented that music has played a significant role in numerous human societies for a long time, but what is this thing called music?

THE VITAL ELEMENTS OF MUSIC

Aaron Copland was an outstanding composer who, more than any other, defined the American sound in classical music. He also wrote a great deal about music because he wanted his listening public to be knowledgeable about the subject. His book *What to Listen for in Music* (1939) became a bestseller. In it, Copland identified what he considered to be the vital elements of music: rhythm, melody, harmony, and timbre. He maintained that the first two—rhythm and melody—are the natural, or instinctive, elements of music.

Rhythm

According to Copland, "If music started anywhere, it started with the beating of a rhythm . . . so immediate and direct in its effect upon us that we instinctively feel its primal origin." He went on to state, "The natural tie-up between bodily movement and basic rhythms are further proof . . . that rhythm is the first of the musical elements."[18] The renowned composer, conductor, and musical theater director Roy Rogosin takes things a step further, asserting, "Rhythm is an organic and necessary part of being alive."[19] The celebrated neurologist Oliver Sacks, testifying before the US Senate, stated, "There is in health an implicit music . . . that is rooted first and foremost in the rhythmicity of all nervous [system] action."[20]

Broadly speaking, rhythm refers to the timing elements of music. In scientific terms, music is time-constrained because it both requires and conforms to the component of timing.[21] Every note is assigned a time value (duration). Even each rest is defined by its time value. Time values also relate the notes to one another: each note lasts a specified length of time, and a sequence of notes adheres to a prescribed length of time. To qualify as rhythm requires a series of notes obeying the factor of time; a single tone held for a defined unit of time is not sufficient.

An important component of rhythm is meter, the repetitive pattern of strong and weak emphasis placed on the notes and rests. Meters can be straightforward, such as a march, 1-2-1-2-1-2-1-2 (2/4 time on sheet music), or a waltz, 1-2-3-1-2-3-1-2-3 (3/4 time on sheet music), or they can be variable, such as 1-2-3-1-2-1-2-3-1-2 (5/4 time on sheet music). The brain detects meters by the recurring pattern of strong and weak emphases. For example, a waltz rhythm has a 3/4 meter and a pattern of strong-weak-weak-strong-weak-weak. Playing a piece of music properly requires consistently maintaining the beat, the underlying pattern (or pulse) of the meter. The brain is a pattern-recognition detector by nature, which is why we tend to find musical pieces without a metrical pattern not only arrhythmic (lacking rhythm), but also difficult, if not downright unpleasant, to listen to.

Another aspect of rhythm is *tempo*, the speed of a piece of music. The tempo can be constant, or it can vary. Slight variations of speed might make a piece more interesting. This is known as *rubato*—literally, stolen time—subtle variations of pace making parts of a piece of music a little slower here, a little faster there.

Rhythm often produces a sense of movement of the body. The art form most closely associated with this is dance. In many styles of dance, people match their bodily movements to an external rhythm. Groups of people can move synchronously by concurrently matching the movement of their bodies to the rhythm.[22] Matching bodily movement to an external rhythm is called entrainment.

Melody

"Melody," Copland wrote, "is second only in importance to rhythm in the musical firmament."[23] He went on to state that if rhythm is connected with physical motion, melody is associated with mental emotion: "Its [Melody's] expressive quality must be such as will arouse an emotional response in the listener."[24]

Melody comprises a series of pitches that form a cohesive whole, the melodic line.[25] The palette of notes with which the composer creates a piece of music is called a scale. How the composer uses the notes on his palette constitutes the art of his craft. He varies the notes—just repeating the same pitch is boring—without making them sound disorganized. "Always remember that in listening to a piece of music you must hang on to the melodic line," wrote Copland.[26] The celebrated conductor and composer Leonard Bernstein, who shared a longtime friendship with Copland, realized Copland was losing his skill as a composer when Copland could no longer maintain the "long line" of his melodies.[27]

On the television game show *Name That Tune*, which debuted in 1952, contestants wagered how many notes they needed to hear to identify a song's title. The contestant who wagered the fewest then had to guess the title after listening to that number of notes. This demonstrates how melody is the foremost aspect in recognizing a song; only a few musical pieces have a rhythm so unique that it identifies them. More than anything else, the melodic line lends a piece of music its identity.

Melody is paramount for a songwriter. Copland related, "A composer is forever accepting or rejecting melodies that come to him spontaneously. In no other department of composition is he forced to rely to the same extent on his musical instinct for guidance."[28] Some melody lines seem to appeal to us intuitively. Indeed, the early great tunesmiths of Broadway felt that their show was a failure unless the audience left the theater singing, humming, or whistling the melodies that they had created. What the critics wrote about the show was less meaningful to them.

Rhythm and melody constitute the core elements, the absolutes of music; there is no music without them. Two other elements which Copland also mentioned are harmony and timbre.

Harmony

Harmony simply refers to all of the notes played that are not melody notes. It is based on two concepts: the interval (how far a note is from the melody note) and the chord (the root note plus two or more notes). Major chords consist of three notes (triads) based on the intervals 1-3-5. For example, the chord of C major has C as its root note, E as its third, and G as its fifth. The simplest harmony consists of playing the root note in the background as the melody line continues on its course. This approach, called a drone, may be how harmony began. A slightly more complex harmony would have instruments play (or voices sing) E or G as harmony notes while C is played (or sung) as a melody note. One way to make this a bit more appealing could be to include instruments or voices that play harmony notes in different octaves; this explains why listening to a female singer accompanied by a cello, which plays in a lower register, tends to be pleasant. Another level of harmonic development would be to add more intervals to the chord, such as the 7 (the note B in the case of C major) to the 1-3-5. Yet another way could be to vary the timing of the harmony notes, for example, by playing them sequentially rather than all at once.

A different approach holds sway in baroque music. Called counterpoint, it entails composing in such a way that each instrument's melody line correlates harmonically with the melody line of the other instruments. This is why baroque music can be quite complex. Some people enjoy this multilayered sound while others do not. A legacy of counterpoint in vocal music is two people singing different melody lines that simultaneously harmonize with one another.

The possibilities for organizing and arranging harmonic note combinations are vast, but the majority view regarding harmony is that it should complement, or even enhance, the melody. Since different traditions of music employ different intervals and organize chords in their own ways, harmony is a major ingredient for giving a piece of music its style and defining its genre.[29] One can take a melody from one genre, say classical, and reharmonize it into a jazzy piece while maintaining its identifiable melodic line.[30]

Copland asserted that harmony is the most "sophisticated" element of music. He further described harmony as an "intellectual conception."[31] By contrast, rhythm and melody come naturally to humans. Rhythm and melody are ancient, prehistoric even, while harmony is a relatively recent, human innovation, developing around the ninth century.[32] While magnificent, harmony is not a natural or instinctive element of music.

Timbre

Timbre refers to the uniqueness of the sound of each instrument, including each singer's voice. Singers refer to their voice as their instrument. Timbre is what makes each voice distinguishable from every other one and what makes an oboe sound like an oboe and not like a violin. Timbre is also called tone color because, in Copland's words, "Timbre in music is analogous to color in painting."[33]

Overtones constitute the crux of timbre. Although a musical note is thought of as a single frequency, the sound produced by an instrument, including the voice, actually contains numerous frequencies. Each note has a base (or fundamental) frequency that is accompanied by other tones occurring at integer multiples above (or "over") the fundamental frequency. Each instrument's profile of overtones is unique, thus imparting its timbre.[34]

Another component that contributes to tone color is how sound waves produced by an instrument interact with their immediate environment. This is called acoustics. Sound waves may bounce off the materials with which they come in contact or, in some measure, be absorbed by them. Every material has a unique profile of how it interacts with sound waves. For example, sound waves bounce off stone while fabric curtains absorb them.

An experienced arranger knows how to combine various instruments to achieve the tone color being sought just as an experienced acoustical engineer knows how to work with the materials in the environment to achieve the desired quality of sound. The properties of the physical materials combine to form the music's tone color and contribute significantly to the listener's experience of the music. Thus, timbre can be viewed as a property of the physical materials used to make music, rather than a property of music itself. Moreover, even a single note played in isolation has timbre, whereas one note in isolation cannot be music.

INTRODUCTION TO THE BRAIN SCIENCE OF MUSIC

Now comes the heady part. Music is not something outside of us. It's not over the rainbow or beyond the sea somewhere. It is within us. By and large, the elements of music that each person receives from "out there" in the world are but the movement of air molecules striking the ear drums. These sound waves then get processed, this input turned into music "in here," in the brain. So, the answer to the question What is music? is, to a great extent, What the brain makes of it.

The three-layer cake brain

Examining the brain's structure is the best place to begin to comprehend how it works its wonders with respect to musical sound wave information. Figure 1.1 shows the major divisions of the brain seen when viewing its surface.

Our primary interest, however, is to take a look inside the brain. A streamlined way to do this is to imagine the brain's structure as a three-layer cake.[35] This is formally known as the Triune brain, shown in figure 1.2. It is a useful, albeit highly simplified, model for thinking about the brain's internal arrangement.

The lowest tier, and the oldest in terms of evolution, corresponds primarily to the brainstem. This layer is tasked with biological regulation, that is, maintaining the body's internal environment. In other words, it oversees the body's housekeeping functions. This layer connects directly to the body's internal or-

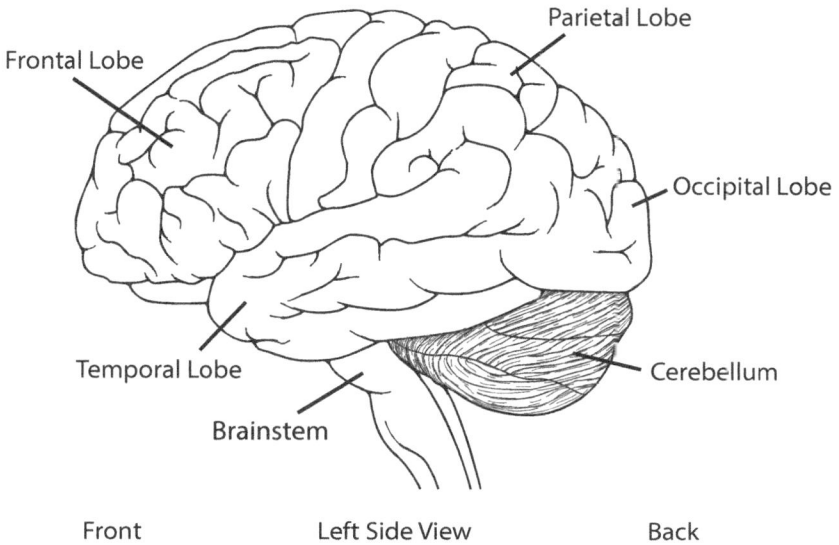

FIGURE 1.1 Lobes of the neocortex along with the cerebellum and brainstem.

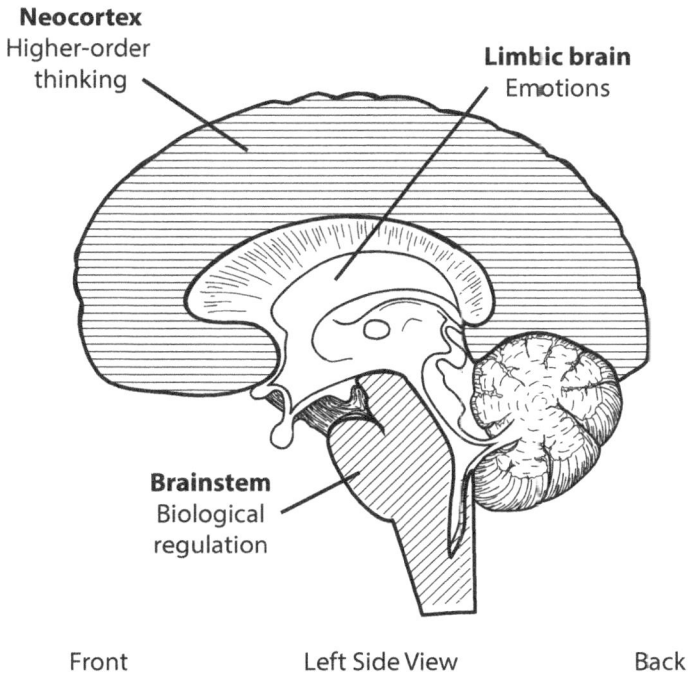

FIGURE 1.2 The triune (three-layer cake) brain and the function of each level. The layers are in frequent communication with one another even though they appear to be separate.

gans via the autonomic nervous system and indirectly via chemical messages coursing through the bloodstream.

The middle tier, the limbic brain, contains the components of the limbic system. Certain forms of memory are located here: episodic (details of our personal life story) and semantic (general knowledge of the world around us). This layer is also the seat of emotions and motivations, both words derived from the Latin verb *movere* (to move). One of the brain's most important movement considerations is whether to approach (move toward/get more of) or to avoid (move away from/get less of) something or someone in the environment. The approach-avoidance calculation is of utmost importance with respect to successful survival and procreation, the principal objectives of evolution.

To give a sense of how long it took for the brain to evolve, humans belong to the order of the animal kingdom called Chordata, meaning animals with a backbone. Chordates first appeared roughly between 500 million and 600 million years ago. The lowest tier of the brain is found in all members of the Chordata order, beginning with fish at the bottom of the order. Meanwhile, the middle tier, the limbic brain, is present in all chordates belonging to the Mammalia class of animals. Mammals first appeared approximately 150 million to 200 million years ago.

Primates first appeared about 6 million years ago. All of them have a neocortex, the uppermost tier of the brain. The neocortex expanded in size spectacularly after humans arrived on the primate scene, perhaps as far back as 2 million years ago. The human brain's highest functions, such as reasoning, abstraction, and imagination, are properties of this layer.

The neocortex is organized into primary and association zones, which can be sensory (input) or motor (output) in nature.

The primary zones interface directly with the outside world, meaning everything beyond the brain, including the rest of the body. Before being sent to a primary zone, motor information first passes through one or more association zones for processing. Likewise, after being received by a primary zone, sensory information is processed by one or more association zones. Figure 1.3 shows the major primary zones. They are found on both sides of the brain, although only shown on the left side in the figure. Association zones, not shown in this figure, are typically located adjacent to their corresponding primary zones.

The primary motor cortex—the physical brain tissue that constitutes the primary motor zone—is located in the rearmost portion of the frontal lobe. It sends instructions to all parts of

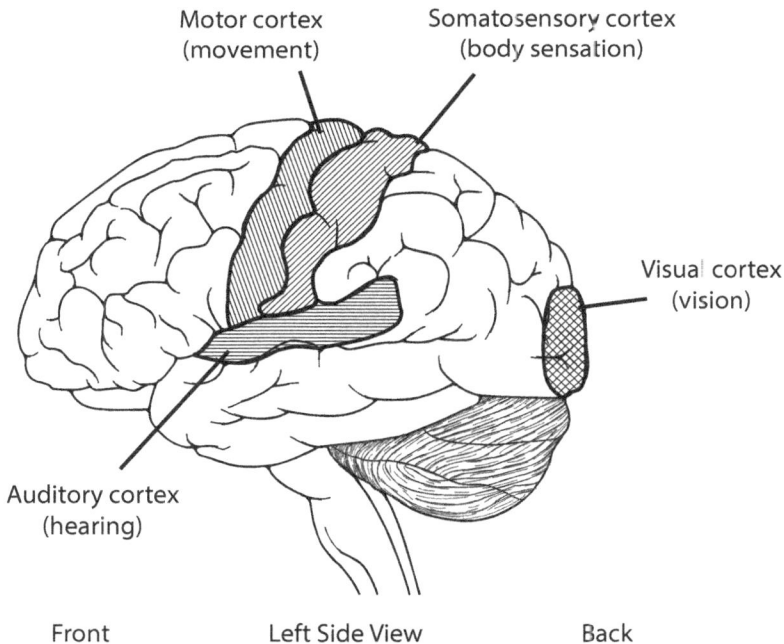

FIGURE 1.3   The major primary zones of the neocortex and their functions.

the body that can move voluntarily, that is, make self-directed movements. Immediately behind this, in the forward-most portion of the parietal lobe, is the primary somatosensory cortex. Somatosensory refers to sensations of the body. Via its many nerves, the body sends information to the brain about the position and motion of all body parts as well as news about anything that comes in contact with the body. **Proprioception** is the technical term for sensing the position and motion of body parts. Dedicated areas of the brain also exist to receive information that pertains to seeing and hearing: the visual cortex is located in the back of the brain in the occipital lobe, and the auditory cortex is located in the superior (meaning "upper") temporal **gyrus**, a fold or bulge of the brain in the temporal lobe.

Approaches to studying the brain science of music

The final item to consider before diving into the brain science of music is how to study it. This may sound a bit philosophical but it's significant. Even when analyzing scientific data, it's important to think about the mindset a person brings to the evaluation of that information. For example, when learning about the basics of brain music science—in particular, when exposed to research showing the large swaths of brain activated by music—it's natural to conclude that vast amounts of the brain participate in converting information about sound waves into music. There's nothing incorrect about that conclusion, but it overlooks a worthwhile question: Are all of the areas that turn on for a given music function, such as melody, required for the task, or are only some of these areas truly necessary? The perspective of the first part of that question casts a wide net, maximizing the impression of how much of the brain music activates. The second part of the question holds a view that is selective, minimizing the sense about which areas of the brain are compulsory for music. It is possible to approach the study of brain music science from either perspective. In fact, even if they seem contradictory, both viewpoints—labeled "maximalist" and "minimalist," respectively—are helpful for gaining a full understating of the subject.[36] It is important to keep both in mind.

Music neuroscientists—more formally, cognitive neuroscientists working in music—are often, though not always, musicians themselves. They tend to use a maximalist approach, seeking to include as many brain regions involved in processing music as possible. One method they commonly employ involves identifying which aspects of sound waves activate which regions of the brain. Their studies of normal brains—sometimes of highly trained musicians, sometimes of novices—have played a pivotal role in revealing how the brain processes various elements of music. A vignette below in the discussion of Sentiments gives an idea of this expansive approach drawing on work with normal subjects

Neurologists—physicians who study and manage disorders of the nervous system, particularly of the brain—are trained to treat patients and to view the brain from the perspective of what disturbance of function results when one or another part of the brain suffers damage. This method of viewing the brain has also played a pivotal role in revealing how the brain processes the elements of music. Neurologists are taught to take a selective, minimalist approach, limiting the processing of music to those areas of the brain that are absolutely required. A vignette in chapter 3, drawing on work with neurological illness, illustrates the approaches neurologists bring to questions on music and the brain.

Both the maximalist and the minimalist methods are valid as well as vital. Their dissimilar orientations raise different questions, generating dialogue that plays a key role in propel-ling the brain science of music forward.

DIVING INTO THE BRAIN SCIENCE OF MUSIC

The time has arrived to immerse ourselves in the brain science of music. Out in the environment, music is but sound waves—vibrations of air molecules. From the human perspective, how-ever, such waves alone do not constitute music. Rather, the brain fashions the vibrating molecules striking the ear drums into the elements of music. The journey begins with rhythm and melody, music's core elements.

Fashioning rhythm

Sound waves reaching the ears are funneled to the cochlea, the hearing receptor organ located in each inner ear. The cochlea converts sound waves from the outside world into electrochem-ical signals that the nervous system can use. These signals exit the inner ear, then pass through several relay stations as they ascend the brainstem. These stations add information about the relative timing of the sounds because the brainstem re-ceives input from both ears, whereas each cochlea receives in-put from only one ear. An example of the utility of receiving in-put from both ears is the ability to locate the source of a sound, since a sound made by an object off to the right will reach the right ear sooner than the left. The operations performed in the brainstem occur automatically and rapidly.[37]

The signals then enter the neocortex at the primary auditory cortex, an area of the upper portion of each temporal lobe. This is the hearing receptor region of the brain. The brain needs to do more than receive hearing signals, however. It also needs to per-

ceive them. This is a multistep process with the next phase occurring in the association auditory cortex. This region, adjacent to the primary auditory cortex, sorts the hearing signals
and shares them with other parts of the brain. The brain can
then perform various assessments on the signals it has received.

The auditory cortex conveys information to the brain's movement and proprioception areas that contribute to perceiving
rhythm. Of note, activation of motor areas occurs for rhythm
perception as well as for rhythm production. It is logical that
even simple rhythm production—say, tapping your finger every
half second[38]—activates the brain's motor system. Interestingly,
this is also the case even if a person remains still, passively listening to music.[39] This observation reveals that rhythm is based
on close communication between the sensory (auditory) and
movement (motor) regions in the brain.[40]

By definition, rhythm encompasses the timing components
of music. Although the brain receives raw timing data about
sound waves from the world, more is needed. Specifically, the
brain has to construct a mental concept of time in order to utilize the raw timing input and formulate timing-related output.[41]
The brain performs this operation in the parietal lobe, its largest sensory region.

The next stage is to reproduce the rhythm that has been perceived. This requires transforming auditory perception of the
rhythm into a motor sequence. This underscores the value of
engaging the motor system earlier in the perception phase: the
system is already in the loop and primed to act. The decision to
move reaches the association motor cortex, regions that function as motor command staging areas. Located just forward of
the primary motor cortex, these areas formulate and communicate specific instructions to the primary motor cortex about

the movements to be made. When the movement actually occurs, it indicates that these instructions have been transmitted by the primary motor cortex to the moving parts of the body.

All movements of the body activate the somatosensory cortex in its role of sensing body position and motion. One of its functions in the case of self-directed movements is to monitor the movements to ensure they follow the specified instructions accurately. Thus, the somatosensory cortex provides real-time feedback to the motor cortex so the latter can make "mid-course corrections" of the movements if errors occur. This keeps us on the beat!

In summary, then, the brain does not possess a dedicated "rhythm center." A multistep operation such as rhythm assessment and reproduction "is not controlled by a single brain region, but by a network of regions that control [various] parameters "[42] of the operation. Rhythm perception comes about through a close interplay between sensation—primarily auditory[43]—and movement regions on both sides of the brain.

Fashioning melody

As with rhythm, melody begins as sound waves entering the ear and funneled to the cochlea. There they are converted into electrochemical signals that the nervous system can use. The cochlea is sensitive to the various frequencies of the sound waves it receives, which is important because each pitch, or musical note, corresponds to a particular sound wave frequency. Music would not be possible without humans' ability to distinguish the various sound wave frequencies that reach our ears.

The electrochemical signals that exit the cochlea maintain their sound wave frequency correspondence throughout their ascent up the brainstem. They then reach the primary auditory cortex, which processes these signals in their preserved sound-wave-frequency organization. Subsequently, this information gets passed on to the appropriate brain regions for analysis.

The bulk of melody assessment takes place in the association auditory cortex. Two critical components of this assessment are relative pitch discrimination and next pitch prediction.[44] Absolute pitch, the ability to hear a single note in isolation and identify it—Aah, that's an F# (F sharp)—is rare.[45] Far more common and important is relative pitch discrimination, which entails recognizing the difference in pitch between successive tones, or what many people typically think of as hearing the difference between two notes. Humans are born with this ability. Discerning the relative interval between notes is paramount for perceiving melody since melody comprises a sequence of discrete and varying pitches that can begin on any note.[46] Next pitch prediction refers to forecasting what tone or tones will come next. Humans are born with the capacity to make next pitch predictions.[47] As one goes through life, these predictions become influenced by one's knowledge of the music being heard as well as by one's personal taste.

Several types of memory contribute to melody assessment. Since melody comprises a string of varying notes in a series, working memory is needed to keep the entire sequence of tones in mind while making relative pitch determinations about the melody's notes. Relative pitch discrimination particularly engages semantic memory, general knowledge about music that is not dependent upon one's personal experience. Next pitch prediction also engages episodic memory, recall of one's personal experiences and preferences. Finally, the brain draws on its recollection of melodies—its library of melody memories—each time it listens to a tune. More than any other factor, melody establishes the identity of a piece of music, so the brain identifies and categorizes pieces of music principally by their melody.

In sum, the brain has no dedicated "melody center." Instead, the multistep operation of melody perception and recognition requires a network of regions to handle the various steps. This highlights a great quality of the human brain—that its many regions communicate effectively with one another. So effectively, in fact, that people experience their brain's functioning as a unified whole even though its actions are divided into numerous components occurring simultaneously in multiple locations that are often distant from one another.

Sentiments

Sentiments are another feature that merits attention in a brain music science primer. *Sentiments* means the combination of emotions and feelings. It is well known that music has the ability to arouse them. Emotions and feelings are closely related to, but not the same as, one another. Nonetheless, they are discussed as one and the same while focusing on music's reward value to the brain here (their differences will be addressed in chapter 2). Music's capacity to reward the brain is the key to how music evokes emotions and feelings.

How does the brain convert signals about the timing and frequencies of sound waves into sentiments? The attributes of music that one values inform a particular brain system about which musical features to reward. This system, located in the limbic brain, is called the reward prediction error system (RPE, or reward system, for short). The RPE responds to predictions about the musical features valued for two key reasons: the high degree of communication (termed *connectivity*) between the auditory cortex and the reward system, and the human brain's capacity to imagine the near future, its ability to predict the next pitch being an expression of this.

As the brain listens to music, the auditory regions share predictions about upcoming sounds with the reward system. The RPE becomes active if there is a mismatch of the actual next sounds with what was predicted. This, in turn, generates sentiments. Sounds simple, yet it is a process built upon sophisticated brain functioning. This phenomenon is probably best communicated by music neuroscientists, who do fabulous work delineating how the normal brain assesses music's components. Finding answers to the primordial question of why we like music is another issue about which these scientists have provided a great deal of information. While taking in the following hypothetical lecture by an archetypal music neuroscientist to a new class of music performance students, bear in mind that the reward system exists in the limbic brain, music highly connects to it, and its activation generates sentiments.

The Sentimental Features of Music

Last week, I spoke about how widely rhythm and melody stimulate the brain. Today's topic is the sentimental features of music. By sentiments, I mean what's typically referred to as emotions and feelings. Back in the 1950s, two scientists, James Olds and Peter Milner, implanted electrodes in various parts of rat brains. They observed that when the electrodes were implanted in specific areas, rats would repeatedly approach and press a lever in order to receive an electrical stimulation. Olds and Milner had discovered a physical correlate for reward in the brain. Analysis of their data showed that these areas are located in the limbic system,[48] sometimes called the "emotional brain," in a region rich in brain cells stimulated by the neurotransmitter dopamine. A neurotransmitter is a chemical released by a cell in the nervous system that signals information to one or more other cells within the system.

Further research has demonstrated that when the brain detects a reward, food for example, it makes a prediction about the value of that reward. If the reward turns out to be better than expected, secretion of dopamine is increased; if the reward is less than expected, dopamine secretion decreases; if the reward matches the prediction, then dopamine secretion remains where it is. The difference between the predicted value of a reward and its actual experienced value is called the reward prediction error. Perhaps "mismatch" would have been a better choice, but "error" is the word we have. Reward prediction errors form the basis of learning about rewards. They also make us strive for more rewards which is usually a beneficial trait from the perspectives of survival and procreation.

Already you can intuit that your task as a performer is to give the members of your audience greater rewards than they expected. But that leads us to an even more foundational question: in what ways does the brain find music rewarding? We can readily understand that valuing food or intimate companionship was already hard-wired into primitive brains many millions of years ago, long before humans appeared on the planet. And it's not hard to imagine that the brain can learn to connect fungible things to rewards, such as money which can be used to acquire a primary reward like food. But music? It's entirely abstract, it's a far more recent development in the history of life on Earth than seeking food or sex, and it's not readily fungible. So, why does our brain bother to like music at all?

The human brain operates by being forward-looking—its time-frame focus is typically "what's going to happen next?"—and it routinely makes predictions about all sorts of things. In fact, making predictions is one of the brain's principal roles. An organism can make a more effective response to an event if that event can be predicted.[49] Robert Zatorre, one of the leading researchers on the reward aspects of music, points out that every musical system has a set of rules governing which sounds follow other sounds. The brain readily learns these rules, called syntax. In fact, children can identify notes that are out-of-key for their culture's music. Thus, we learn that, from a young age, the brain is busy making predictions in real time about the music that it's hearing.

But how does music connect to reward? Well, it turns out that the brain's sensory systems, the auditory (hearing) system in the case of music, communicate with the reward system. When the brain hears a sound, it makes a prediction about the next sound. If the next sound turns out to be better than what was predicted, the reward system will respond with increased dopamine release, the currency of reward in the brain. Dopamine release triggers sentiments (emotions and feelings) associated with reward that are commonly called pleasure. Interestingly, about 5% of the human population does not derive pleasure from music. The technical term for this is "music anhedonia," and studies have shown that the auditory system is not well coupled with the reward system in these individuals.[50]

Which sounds any one person likes to hear is determined by a combination of familiarity with the rules of one's musical culture along with learning, personal exposure and individual taste. There's no way everyone will want to listen to your music: some people connect with your genre of music and some people don't. That's just the way it is. Your job as a performer is to guide people who are open to your manner of music through a journey of auditory experiences that activate wide expanses of their brains and auditory expectations that stimulate their reward system. The task before each of you here is to develop your own personal style to bring your listeners on that journey with you. I wish you all the best in your endeavors.

Musical Memory

The power to evoke autobiographical memories represents another connection between music and sentiments. For example, Jean-Joseph Mouret's Fanfare-Rondeau from Suite of Symphonies No. 1 for brass, strings, and timpani became a successful theme song for the television series *Masterpiece Theater*. Whenever I hear this piece, I automatically associate it with a specific *Masterpiece* series, *The Six Wives of Henry VIII*. I was a child at the time, and hearing Mouret's music vividly brings my father to mind. An avid English history buff, my dad adored the show.

A few years ago, I attended an art exhibit on the Tudor dynasty. There on the wall hung a portrait of Henry VII, the dynasty's founder. Farther along was Holbein's famous portrait of Henry VIII (1536–37), striking an akimbo pose and wearing a wide-brimmed hat.[51] Next, Elizabeth I, posed in a manner imparting an air of regal strength. Finally, Sir Thomas More, the man for all seasons, whose principles cost him his head. Mouret's Fanfare played between my ears the entire time as I roamed the exhibition, effortlessly recalling episodes from *Masterpiece Theater*. And, oh, how I wished my dad, who died more than forty years earlier, could have seen the exhibit with me.

This type of memory has a name: music-evoked autobiographical memory (MEAM). A person hears a certain song and—Boom!—they're suddenly back in the past, recalling the event not in a detached manner but emotionally reliving it. How is it that music can evoke strong reminiscences and sentiments from the past? It begins with encoding, the process of forming memories, and continues with music's connection to the emotional brain.

Songs are best encoded when associated with strong emotions and feelings. This explains why people most easily recall songs associated with emotionally charged times in their lives. These memories can then lay dormant for long stretches of time until hearing the music reawakens them, for music has the ability to spark memory involuntarily. No need for mental stressing and straining to remember since the prompt arrives from outside ourselves. And, in such instances, the piece of music not only recalls memories, it rekindles the accompanying sentiments as well. This is because the external cue feeds directly into the limbic brain, which explains why emotions and feelings accompany the reawakened memory.[52]

WHEN I SAY "MUSIC," I MEAN . . .

This book communicates certain key concepts through the word "music." The first is a fundamental definition of music that is expressly streamlined to optimize clarity and ease access to active musical engagement. As already noted, rhythm and melody comprise the essential components needed to meet the definition of music. Together, rhythm and melody are music in its most basic form, its core elements.

A grand orchestra performing a glorious symphony, a large ensemble striking up a rousing march, a solo piano playing a

A Word from a Music Neuroscientist

One of the joys in writing this book has been the people I've had the privilege and opportunity to meet along the journey. Hervé Platel, a professor at the University of Caen, Normandy, France, is an internationally respected music neuroscientist with several decades of experience. His varied and impressive research interests range from the theoretical to the therapeutic.[53] He was kind enough to share some of his thoughts about his field, both its history as well as future directions.

When we started using revolutionary neuroimaging techniques to explore the musical brain in the 1990s, we naively hoped to find a brain center for music. It was a great surprise to observe that simply listening to music provokes a veritable "neural symphony" in the brain. This opus involves not only the neural networks involved in perception, but also in the processing of emotions, memory, and even motor skills.

What might have seemed like a disappointing message—there is no cerebral center for music—turns out to be particularly important: music stimulates our brain very powerfully and very broadly. This is why we can make use of music to stimulate many types of underperforming functions (language, motor skills, memory, emotional regulation, attentional capacities, etc.) in patients with brain illness. Work in the cognitive neuroscience of music carried out since the beginning of the 2000s has now made it possible to identify the mechanisms that explain why, and above all how, music can be therapeutic.

Is this the end of the story? Certainly not. New frontiers to be explored now stand before us. These will lead to better understanding of the effects of neuroplasticity induced by the simple listening of music as well as music's impact on the very expression of our genes, showing that musical stimuli can modulate gene expression.[54] Musical epigenetics is an incredibly vast field of exploration, which opens up amazing perspectives on our understanding of the therapeutic power of music.

Professor Hervé Platel
University of Caen
July 2024

luscious sonata, a multiple-piece band laying down some driving rock 'n roll—all surpass the essence of music. Think of viewing a complete window treatment. There is molding, screens, blinds, drapes with swags and jabots, and so on, but what is the essence of window? Likewise with music, the entire treatment makes for a wonderful experience. The question remains, however, what is the essence of music?

The core elements

Rhythm, encompassing aspects of timing, is a core requirement for anything to be considered music. Tones that occur randomly in time, without any concern for when they occur or how long they last, flunk the music definition test. Similarly, a single tone in isolation has no rhythm because rhythm requires a relationship of tones with one another.[55] Some rhythms are simple and steady, others compound and variable. Regardless of the level of complexity, an underlying timing structure must be present to pass the music sniff test. The brain, as a pattern-detection machine, will not perceive a series of tones lacking a rhythmic pattern as musical.[56]

The other essential of music, melody, consists of a sequence of pitches that forms a coherent whole—the melodic line. Like rhythm, melody entails a relationship among notes, so an isolated tone cannot be considered a melody. More than any other ingredient, melody imparts identity to a piece of music.

And there you have it: At its definitional essence, music refers to a coherent series of pitches (melody) that conform to time conditions (rhythm).

Harmony adds layers of complexity to music, principally as an ingredient that connects a piece of music to a particular style, or genre, of music. Wonderful to experience and appreciate as a listener, harmony allows those with expertise in music composition to excel. As stated earlier, it is (relative to rhythm and melody) a newcomer to the musical scene and an intellectual innovation. By contrast, the core elements of music, rhythm and melody, come naturally to humans. Thus, harmony enriches the experience of music, but is not a core element of music.

Timbre is a physical property of an object producing a sound. Any physical material making a sound possesses timbre since sound waves exist in the physical world. The timbre of a certain object may not be pleasing, but that does not nullify the fact that it possesses timbre. Additionally, even an individual sound in isolation possesses timbre whereas music requires a series of sounds. Timbre is, therefore, a core element of sound rather than of music and is found in all music by virtue of its applying to all sound.

The human body produces sounds of numerous timbres. The voice is the greatest source, but other body parts can as well, such as the hands (clapping) and feet (tapping). Instruments add richness and complexity to music, but the human body alone is sufficient to make music. Put another way, not only can the body receive a series of pitches abiding by time conditions that the brain can turn into music, it is also capable of producing them.

What's "out there" and what's "in here"?

The second key concept of "music" involves the distinction between what's "out there" and what's "in here"—that is, what's outside the brain and what's inside the brain, respectively. What's out there are sound waves, vibration of air molecules, that reach the ears, which convert them into electrochemical impulses that travel in here, to the brain. Ultimately, music is what the brain makes of sound waves turned into electrochemical nervous system impulses.

As for a definition of the brain in the case of music, it is the processor and interpreter of the sound wave information received from the world. The vast enterprise called music results from these brain operations, including but not limited to assessing rhythmic and melodic sound wave information, generating sentiments, learning from and remembering music, and planning and participating in musical output.

There are several approaches to ascertaining how the brain interacts with musical sound wave information. This chapter highlighted the maximalist approach, which emphasizes studying normal individuals and leads to an expansive understanding of the amount of brain tissue devoted to music. The minimalist approach, focusing on examining abnormal brains and tending toward a circumscribed view of the extent of the brain devoted to the music process, was also mentioned and will be discussed in detail in chapter 3. Both viewpoints are vital and valid, and as noted earlier, advocates of the two approaches engage in dialogue that adds to appreciating and understanding the music-brain relationship. The terms *maximalist* and *minimalist* are simply labels used here to differentiate the two perspectives for ease of comprehension.[57]

The bottom line: music is a pas de deux of sound waves, from "out there," with the brain, "in here." The focus of this book is the power of that interplay over the course of evolution and through personal engagement.

Music for the Hearing Impaired

If the "out there" of music is sound waves, where does that leave people who are hearing-impaired? Can deaf people participate with music? As a matter of fact, they can because deaf people have intact vibration sensitivity. Being vibrations of molecules in the air, sound waves can be sensed by vibration receptors on the skin.

Robert Gault proposed the concept of "listening" to music via skin vibrations in the mid-1920s.[58] The technique, termed vibrotactile perception or vibrotactile sensitivity, is based on humans having four types of skin receptor cells for vibration, each with its own sound wave frequency sensitivity range.[59] In England, the University of Liverpool collaborates with the Royal School for the Deaf Derby to study this approach. Their scientific studies encompass the physiology of touch and the physics of structural borne sound. The collaboration has led to the development of equipment to help deaf and hearing impaired people experience musical vibrations on the glabrous (hairless) skin of their hands and feet.

This research demonstrates that the notes detectable by their method range from C_1 to G_5.[60] By comparison, the range of a piano is A_1 to C_8 so their method favors the lower notes within the range that is perceptible by human hearing. The school's research has found that the feet are better than the hands for detecting the lowest notes. The smallest interval of discrimination by the school's method is three semitones, also known as a minor third, slightly less sensitive than individuals affected by what's commonly called tone deafness (see chapter 3).

The bottom line of this collaborative work is that effective spectral (pitch frequency) discrimination can be achieved via vibration, so deaf and hearing impaired people can appreciate —listen to and also play—music via vibration. Their work has even shown that some timbre discrimination is possible, so the unique qualities of different instruments can be appreciated.

For further information, take a moment to view the absorbing video at University of Liverpool, Acoustics Research Unit, "Musical Vibrations: Using Vibrotactile Technology to support d/Deaf People in Music Education at the Royal School for the Deaf Derby" https://stream.liv.ac.uk/2qvwd9th.

CONCLUSION

Main points

- Music is ancient, and its origins are shrouded in the mystery of the past. Humans, at some point in their evolution, gained the ability to make it. When that occurred continues to be debated.

- Music results from the brain's processing and interpretation of sound waves that form a coherent melodic line and adhere to time conditions (scientifically, called time contraints).

- A "three-layer cake" is a useful, albeit simplistic, model for thinking about the brain's structure: higher cognitive functions occur in the upper (neocortex) layer, sentiments originate in the middle (limbic) level, and biological regulation takes place in the lower level (brainstem). The neocortex consists of primary zones (for interfacing with the world outside of the brain) and association zones (for processing and interpreting information).

- Rhythm refers to the timing components of music, requires close auditory-motor communication, and is associated with bodily motion. Melody gives a piece of music its principal identity, relies upon relative pitch discrimination and next-note prediction, and is associated with creative inspiration.

- Music readily elicits sentiments, particularly by means of the rich connectivity of the brain's auditory region with the reward system.

- Music is complex. There is no single "music center" of the brain; rather, perceiving and producing music makes use of many different regions and abilities of the brain.

A few thought-stimulating questions to ponder

- How do you imagine music sounded when humans were first able to make it?
- Is rhythm or melody the more important component of music for you?
- Which instrument timbres and voice ranges do you prefer?
- How do you utilize music to improve your everyday life functioning?

NOTE TO READERS

You now possess enough knowledge about the underlying science to begin exploring music's power to the brain from an evolutionary perspective. Still, there's a lot more to learn about that science and familiarizing yourself with it will add greater depth to your exploration.

There is now a choice to make: Option 1 is to become familiar with the additional background knowledge in chapter 2. This information is fascinating albeit more technical than that in chapter 1. Option 2 is to skip ahead to chapter 3, deferring chapter 2 until later. This route gets to the core material of the book more quickly.

The choice is yours. Pick Option 1 if you're interested in acquiring broader background knowledge before continuing on or if you were already familiar with the information in chapter 1. Alternatively, choose Option 2 to engage with core material now, returning to the additional background info of chapter 2 when you're ready. I do recommend that you come back to it.

There's no right or wrong decision. Either way is valid.

Happy reading.

A Closer Look at a Musical Brain

"Music is a higher revelation than all wisdom and philosophy."
—*Ludwig van Beethoven*

Brain is soft tissue. Unlike bones and teeth, it rapidly decomposes after death, so there simply are no old brains to study. Embalmers preparing mummies in ancient Egypt didn't even bother with the brain; they considered the organ unimportant, unworthy of preservation. Not only is there scant physical evidence of brains from the distant past, until recently scientists had little understanding of how the organ between the ears functions. Little wonder then that ideas and concepts about the brain have varied widely since antiquity.

Misunderstandings about the brain fill the Western literary record, or perhaps it's more accurate to say that an understanding of the brain is often absent from the literary record. The Hebrew Bible makes reference to multiple organs—particularly the heart, but also the kidneys and liver—in connection with thoughts, sentiments, and behavior, but omits the brain. In James Strong's *Exhaustive Concordance of the Bible* (1890)—a listing of primary words in the King James Version of the Bible—the brain receives one mention, whereas the heart is cited 826 times.[1]

The ancient Greeks were divided in their ideas about the brain. On the one hand, Aristotle (384–322 BCE) "saw the heart as the seat of spiritual and mental functions . . . [with] the major task of the brain [being] to cool the heart, which was often too hot-blooded."[2] On the other hand, Hippocrates (460–370 BCE) asserted, "From the brain and from the brain alone arise our pleasures, joys, laughter, and jests, as well as our sorrows, pains, griefs, and tears."[3] Several centuries later, an anatomist named Rufus of Ephesus (80–150 CE) produced a general physical description of the brain.[4] The positions of Hippocrates and Rufus anticipate what would become the modern scientific perspective of the brain.

SCIENTIFIC STUDY OF THE BRAIN

Modern comprehension of the brain's anatomy began to take shape around the middle of the seventeenth century, when the English physician Thomas Willis published an in-depth description of its physical organization.[5] Details about its anatomy and physiology have expanded ever since, calling attention to the interrelationship between the brain's structure and function. Contemplating this connection through the lens of evolution leads to thought-provoking insights about the purpose of the brain.

Brains serve bodies

Brains exist in organisms that are mobile, while life forms rooted in the ground lack brains. This implies that the most fundamental purpose of a brain is to enhance an organism's ability to move about in its environment in order to satisfy its needs. A corollary to this insight deems the brain to be "of the body"— meaning, its primary purpose is to work for the benefit of the body. So, while the human brain possesses many orders of additional complexity, the organ's foundational principle of existence is to serve the needs of the body it occupies.[6]

How did the earliest brains of the animal kingdom communicate with the bodies they served? Consider an analogy using the Pony Express.[7] Organs of the body sent packages of chemical information to the brain via the bloodstream and the brain sent chemical messages of instruction back to the organs by the same route. It was like letters being carried over old Western trails from Missouri to California by horse: quaint and romantic, but slow and unreliable. No wonder the Pony Express lasted only nineteen months before being replaced by trains. Likewise, the arrangement of brains communicating with organs of the body solely via the bloodstream yielded only simple organisms, but the adaptations—in the form of genetic mutations—needed to advance the situation occurred over evolutionary eras, rather than in a year and a half.

Organisms gradually evolved with dedicated neural pathways directly connecting the brain to the viscera, the body's internal organs, and vice versa. These conduits constitute the autonomic nervous system (ANS). Communication along the ANS was more rapid and reliable than along the old bloodstream routes, but two important points should be kept in mind. First, unlike the real-life Pony Express vanishing into the pages of history books, the body's "pony express" approach remained on the scene after the ANS arrived. Indeed, the old bloodstream method is still in use today, hormones being a good example. The process of brain evolution builds a new layer on top of an older layer, and the prior layer continues to function.[8] Second, the brain locates the information it receives from the body as belonging to the body part from which the signal was sent. So, for example, when the brain registers information whose source is the heart, it locates that information as originating in the heart.[9]

The earliest brains attended to basic bodily needs, a process known as biological regulation, corresponding to the lowest tier of the triune brain model (see figure 1.2). These brains had little, if any, direct interaction with the world outside the body. Over the eons of evolution, a higher level of complexity appeared, connected to the development of the middle tier. This tier's motor and behavioral skills enhanced the ability of those who possessed them to survive and reproduce. This accords with the theory of evolution: traits that increase an organism's ability to survive and reproduce are selected for, meaning consistently passed on to future generations.

Middle-tier motor skills resulted in improved coordination of body movements thanks to brain circuitry to optimize common and repetitive movements.[10] Around the same time, the peripheral nerves, another type of conduit, came into being to link the brain directly to every corner of the body, including the muscles used to move the body. An organism better able to move about in its environment is more likely to survive and reproduce.

Another development involved the appearance of sense organs. Rudimentary at first, the brain could now directly access specialized information (such as smell, vision, and hearing) about the world outside the body. This accelerated the process of inputting information to the brain, which could then more quickly output a response message to the body.

Emotions also emerged as the middle tier developed. In fact, the middle tier is, in large part, associated with the limbic system and often called the emotional brain. In neuroscientific terms, emotions are related to, but are not the same as, feelings. Stemming from the word *motion*, emotions are akin to patterns of action involving multiple components of the body capable of working in unison under the brain's baton. Think of an emotion as a sort of "playbook" facilitating all of the members of the team (the parts of the body) to be "on the same page."

The brain's ability to form memories of various sorts also arose with the advent of the middle tier. This major advance boosted the brain's ability to enhance its, and hence the body's, performance through the process of learning. From the perspective of evolution, learning represents a huge leap forward, giving organisms a fresh means to form and retain new behaviors. Organisms able to use memory and learning can adapt to changing environmental conditions much more rapidly than those that require a genetic mutation for improved performance. Species that are better able to learn have a robust arrow in their evolutionary quiver.

The uppermost tier of the brain developed many eons after the middle tier. This level, called the neocortex—literally, the new surface—is divided into four sections or lobes (see figure 1.1). Please note that the word *cortex* does not mean "core," a clang association that can lead to a mistaken understanding of the word. The neocortex comprises numerous functional zones. Primary zones interface directly with information entering (input) and exiting (output) the brain. Association zones process information linked to one or more primary zones and can also draw upon memory as they process information. They are typically located adjacent to the primary zones (see figure 1.3).

As a final point before leaving this section, there is a particularly vital way in which the body serves the brain: providing it constant nourishment. Frankly, the brain is an energy hog that the body has to constantly feed. To accomplish the wonderful things it does, the brain consumes 20–25 percent of the body's oxygen and nutrients.[11] It does this twenty-four hours a day, whether a person is awake or asleep.

LOCALIZATION VERSUS EQUIPOTENTIALITY
OR GENETICS VERSUS LEARNING

The background information just presented about the brain's architecture and reason for being leads to considering how the brain works, especially the neocortex, the most advanced layer and the one that most distinguishes the human brain from the brains of other species. A long-running dispute about how the brain conducts its business has been waged between the proponents of mass action versus those of **localization**. In a nutshell, the former group claims that brain function is determined foremost by how much of the brain is brought to bear on a task (mass action), whereas the latter group asserts that the brain operates by each region performing a specialized function (localization). Mass action thinking persists in everyday speech to this day. The assertion that humans only use 10 percent of their brain is a vestige of the mass action perspective. It seems easy enough to identify the key weakness of the mass action position—if 10 percent of the brain were removed, it wouldn't necessarily lose 10 percent of its function because this would depend on which 10 percent is removed—yet the debate raged for decades.

The **lesion method**, the "gold standard" for determining what a particular area of the brain does, emerged as the key technique for verifying the concept of localization. The term **lesion** is nonspecific, as it refers to any sort of objectively identified abnormality in the brain. A **stroke** represents one such example, while a tumor represents another. According to the lesion method, what role (or roles) a particular area of the brain performs can only be truly known through studying the brain's activity when the functioning of that area is lost or markedly abnormal. Of course, it would be unethical to induce a lesion on purpose to a human being,[12] so physicians and researchers have to wait until a patient with a lesion in a brain area of interest chances to show up.

Just such a patient appeared in the mid-1800s. Louis Victor Leborgne was hospitalized in Paris due to a gradual decline of his ability to speak to the point that he could only utter the sound "Tan," the name often attached to him in the medical literature. Pierre Paul Broca, a French physician, studied Tan late in the patient's life. In 1861 Broca published Tan's autopsy report and previous examination findings about him. The autopsy revealed a lesion in the inferior frontal gyrus of the left frontal lobe, a site now called Broca's area (figure 2.1). More reports of similar patients followed, allowing a connection to be drawn between the patients' difficulty expressing language and a lesion at this particular site in the brain. Score a big point for localiza-

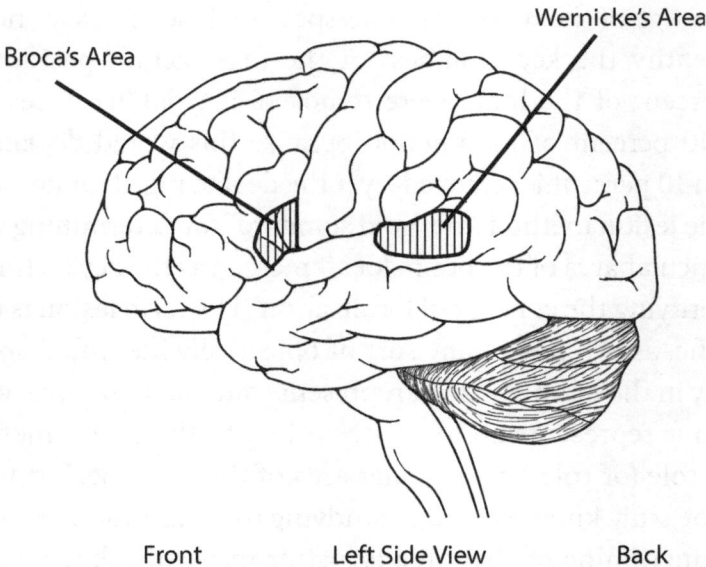

FIGURE 2.1   Language areas identified by and now named after Pierre Paul Broca and Carl Wernicke.

tion. Several decades later, a German physician, Carl Wernicke, presented evidence of a lesion found during an autopsy of another language area of the brain, now called Wernicke's area, that resulted in difficulty with reception (comprehension) of language.

Localization appeared to be on a roll, but the concept of mass action failed to disappear for two major reasons. First, at the time no convincing way existed to confirm that a lesion had actually occurred during a person's life. After all, wasn't it possible that the lesion appeared only after death, perhaps caused by the act of dying or the autopsy process? Second, the notion of mass action evolved into a more refined concept, the equipotential brain or equipotentiality. This concept did not deny that different parts of the brain serve different functions, but claimed that this was due to learning; at birth, all parts of the brain have equal potential to learn whatever function they are taught.

The debate raged for decades. Further autopsy studies continued to reveal correlations between the loss of certain abilities and identifiable brain lesions, but no one could find a specific lesion interrupting such functions as long-term memory or personal identity. The pugilists fought to a draw through the first half of the twentieth century.

Enter the brilliant neurosurgeon Wilder Penfield. While performing operations to control seizures in epileptic patients, Penfield observed the responses to stimulation of the abnormal tissue as well as of nearby normal parts of the brain's surface. Since the brain has no pain receptors, Penfield could perform these operations while the patients were awake. This work led to the landmark 1954 publication of the somatosensory and motor homunculi (figure 2.2).[13] A homunculus (plural homunculi) is a representation, or map, within the brain of the body.[14] Localization—identification of a specific part of the brain

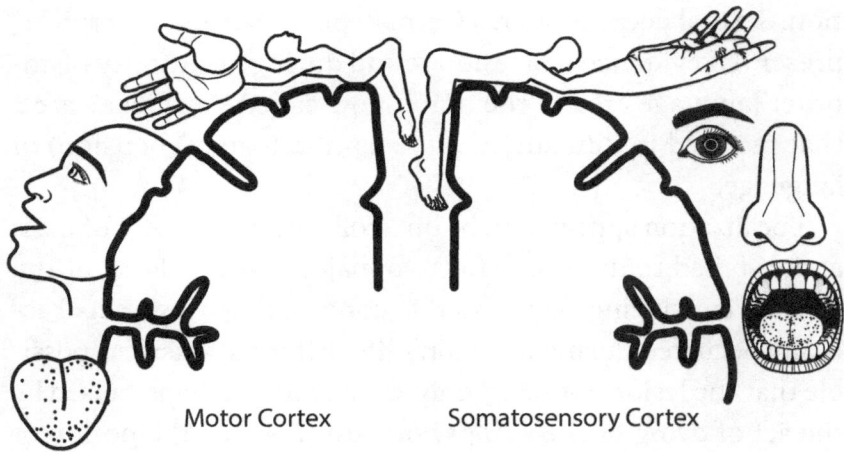

Motor Cortex   Somatosensory Cortex

FIGURE 2.2   Parts of the body served by the motor (*left*) and somatosensory (*right*) homunculi viewed from the front. For the locations of the motor (movement) cortex and somatosensory (body sensation) cortex from the side view, see figure 1.3.

serving a specific function in a living person—appeared to have delivered the knock-out punch.

The second half of the twentieth century brought further confirmation of the principle of localization once localized lesions could be seen literally in real-time thanks to the development of computed tomography (CT) and magnetic resonance (MR) imaging. Later, with the advent of functional scanning, the brain could be imaged as it worked. This type of imaging solidified the concept that brain function is accomplished through the interconnection of numerous functional modules, such as Broca's area and Wernicke's area. These modules are composed of **neurons** (so are located in **gray matter**), work in parallel (hence can operate independently and simultaneously), and are widely distributed across the brain. Tracts—that is, **white matter** pathways such as the arcuate fasciculus—connect the modules to one another. The modules involved and the tracts that connect them are genetically determined and form units called networks. As an example, two modules (Broca's and Wernicke's areas) interconnect via the arcuate fasciculus to form a portion of the language network (Figure 2.3).

"In other words," wrote Michael Gazzaniga, a leading contemporary researcher and author in cognitive neuroscience, "[brain] capacities are genetically determined [by] neural networks [that are] specialized for particular kinds of learning"[15] That is, the general categories of function—such as vision is served by this network of the brain and movement is served by that network of the brain—are determined by our genes.

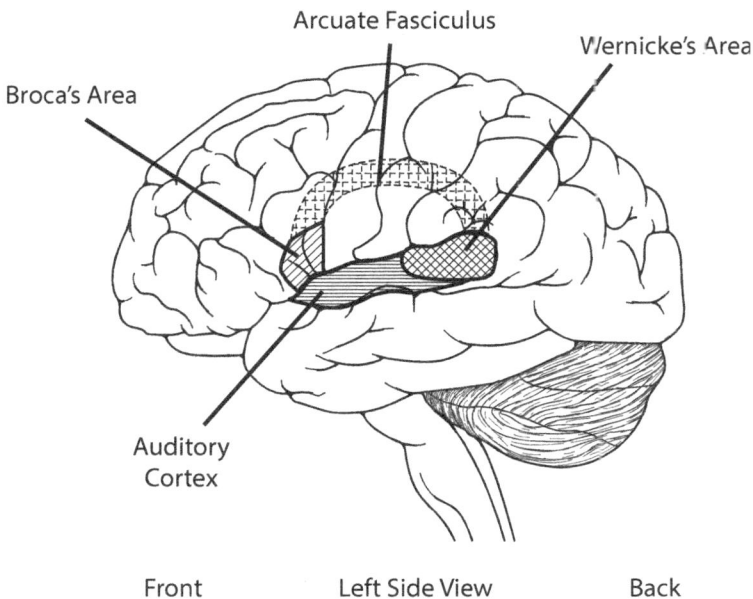

FIGURE 2.3   An example of two modules (Broca and Wernicke's areas) connected via a tract (arcuate fasciculus) in the language network of the left hemisphere.

Still, localization could not explain all brain phenomena. While the naked eye with the aid of a scanner can chart genetically determined networks, it cannot account for what occurs at a cellular level. The phrase "neurons that fire together wire together" was coined to explain neuropsychologist Donald Hebb's description of how the brain learns.[16] Hebb's principle, proposed in the late 1940s, remains a key component of the physiologic basis of memory and a guiding concept of connectivity, how neurons form connections with one another. As Gazzaniga observed, "Hebb's ideas pointed out the centrality of the idea of the importance of connectivity [and] it remains a central topic of study in neuroscience today."[17]

Brain cells fire (activate) when they receive new information. Through repetition, the neurons that fire together in response to the information form permanent connections (wire together), which forms a memory. Lesson learned. Thus, education and experience shape the specific contents of the module. Although each module of the network is designed to operate within a certain category of function, the process of learning makes a wide potential range of output possible. For example, certain regions of the brain are genetically destined to perform language functions (localization), yet a young child has equal potential to learn any human language. Likewise, the brain has innate capacities for music and, thanks to learning, humans have developed many styles of music and can engage with music in different ways.

THE NEOCORTEX AND MUSIC PROCESSING

What does the neocortex do when it comes to processing music? Musical sound waves arrive at the cochlea, the inner ear hearing organ. The cochlea's structure makes it sensitive to the various frequencies of the sound waves it receives, each musical note corresponding to a defined sound wave frequency. The signals then exit the cochlea and enter the nervous system, ascending the brainstem. Timing information about the sounds is added in the brainstem since it receives input from both cochleae.[18] Sensitivity to the various sound wave frequencies is preserved throughout this brainstem ascent. The signals then reach the primary auditory cortex, regions of both temporal lobes specialized in receiving sound signals. Like the cochlea, the primary auditory cortex is **tonotopic**, sensitive to the various sound waves frequencies it receives (figure 2.4).

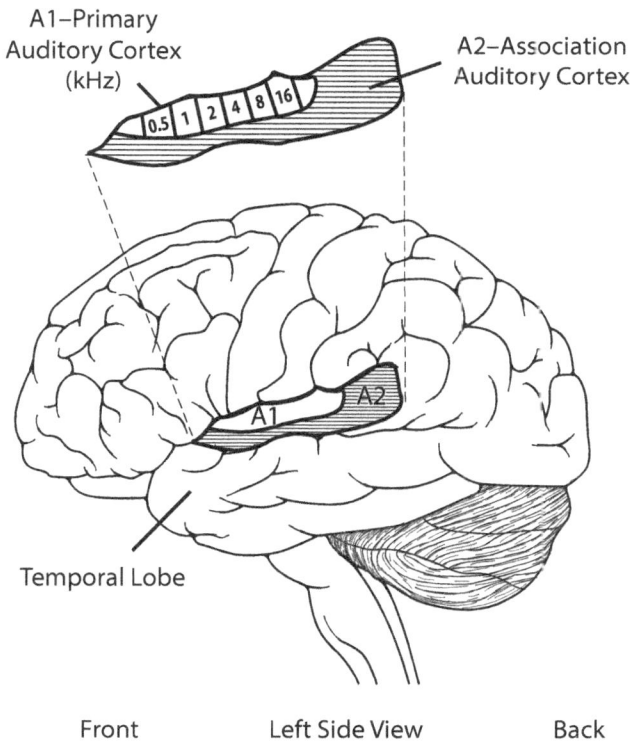

FIGURE 2.4 Location of the auditory cortex in the temporal lobe, along with (*inset*) a tonotopic map of the primary auditory cortex organized by kilohertz (kHz)

At this point, the brain makes a gross determination about what type of sounds it's receiving. Timbre contributes to the brain's assessment of the type of sound waves the ears are hearing by helping to identify the source of the sound. Sounds like a clarinet? It's music. Sounds like my friend's voice? No doubt, it's language. Sounds like two pieces of metal banging against each other? Probably just noise. In the case of music, the primary auditory cortex packages and sends information to the association auditory cortex where analysis of sound wave input and crafting of rhythm and melody continue.

Crafting rhythm

Rhythm perception and production involve a close connection between auditory and motor networks. Studies of rhythm processing several decades ago confirmed the involvement of the association auditory cortex, but there was no consistent localization or lateralization of this activity within the association auditory cortex. (Localization means a function occurs in a unique area of the brain; lateralization means a function occurs exclusively on one side or on one side of the brain more than on the other.) This led to idea that the processing of rhythm "may depend to a large extent on interactions between the auditory and motor systems."[19] Subsequent work established that this interaction occurs even when simply listening to rhythm while lying still.[20]

In 2020 a team of researchers in Japan reported that rhythm stimulates activation on both sides of the brain of multiple motor and sensory system areas in the neocortex (figure 2.5). They attributed five stages to the process by which the brain assesses rhythm:

1. Arrival of sound input in the primary and association auditory regions.

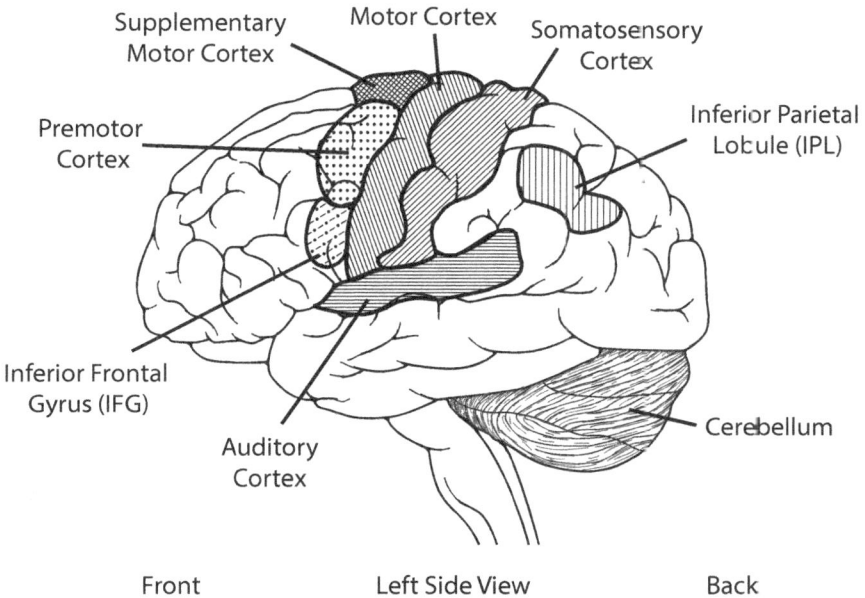

FIGURE 2.5 Regions visible on the left hemisphere of the brain's surface that are active during rhythm assessment.
Adapted from Naho Konoike and Katsuki Nakamura, "Cerebral Substrates for Controlling Rhythmic Movements," *Brain Science* 10, no. 8 (2020), https://doi.org/10.3390/brainsci10080514, with the kind permission of Naho Konoike.

2. Detection of the rhythmic structure activates the inferior frontal gyrus and adjacent premotor cortex plus the inferior parietal lobule.

3. Transformation of the sensory percept of rhythm into a motor percept of movement (even if the body remains perfectly still), a function centered at the inferior parietal lobule (IPL).

4. Initiation of "motor imagery/rehearsal" (as named by the researchers), this stage is associated with activity

in the supplemental motor area (even if the body
does not move).

5. Involvement of the primary motor and somatosensory
cortices when body movement happens, the former
relaying movement instructions to the body, the latter
monitoring the movements in a feedback manner.[21]

Activation of the premotor cortex, even during the perception of rhythm, is consistent with the concept of rhythm stemming from an interaction between the auditory and motor
systems. That is to say, rhythm "is hypothesized to rely on the
integration of sensory [auditory] information with temporal information encoded in motor regions such as the . . . premotor
cortex."[22] It is also consistent with the view that rhythm and
movement are closely intertwined, as the premotor cortex "is
believed to have direct control over the movements of voluntary muscles."[23]

The inferior frontal gyrus plays a major role with respect to
syntax, the rules governing which pitches (for music) and phonemes (for language) follow one another. Syntax has a part to
play in the rhythm domain since the timing of pitches is also
determined by a rule-based system.

A region of the parietal lobe, the IPL, plays a pivotal role in
the Japanese researchers' model during stage 2 and stage 3,
which encompasses the perception of the timing structure of
rhythm and its transformation into movement. The authors observed, "The IPL is involved in representing the temporal information of rhythm."[24] Soundly presenting their case, they cite
supporting data from both functional imaging studies and lesion studies—that is, evidence gathered from both normal and
abnormal brains. They concluded that the IPL is a key component of the brain's capacity to sense and conceptualize time.[25]

The cerebellum and the basal ganglia, regions below the neocortex, also play valuable roles with respect to rhythm. Tucked under the back part (occipital lobe) of the neocortex, the cerebellum is sometimes called the "coordination center" of the brain. In the five-stage rhythm process outlined above, the cerebellum is active during both analysis of an incoming rhythm and planning the motions to produce an outgoing rhythm.[26] Multiple researchers posit that the cerebellum's role involves leveraging incoming sensory information to optimize feedforward control—preparation and anticipation—of outgoing movements (in this case, movements associated with rhythm).[27]

The basal ganglia are collections of brain cells located deep on either side of the midline near the center of the brain that cannot be seen from the brain's surface (hence, their omission from figure 2.5). Researchers propose the involvement of the basal ganglia in controlling rhythmic movements on a millisecond time scale such as those necessary for supervising well-rehearsed motions not requiring conscious attention.[28]

In summary, the above information regarding the brain regions involved in rhythm processing support two key concepts. First, rhythm results from the interaction of multiple brain regions working together; there is no single "rhythm center." Second, rhythm requires a close connection between the auditory and motor networks, for both perception and production.[29]

Crafting melody

After the arrival of sound wave input to the primary auditory cortex, the bulk of melody processing takes place in the association auditory cortex. Additionally, working memory and syntax review, vital frontal lobe functions, are necessary for melody assessment. Figure 2.6 charts key areas of the neocortex that contribute to crafting melody.

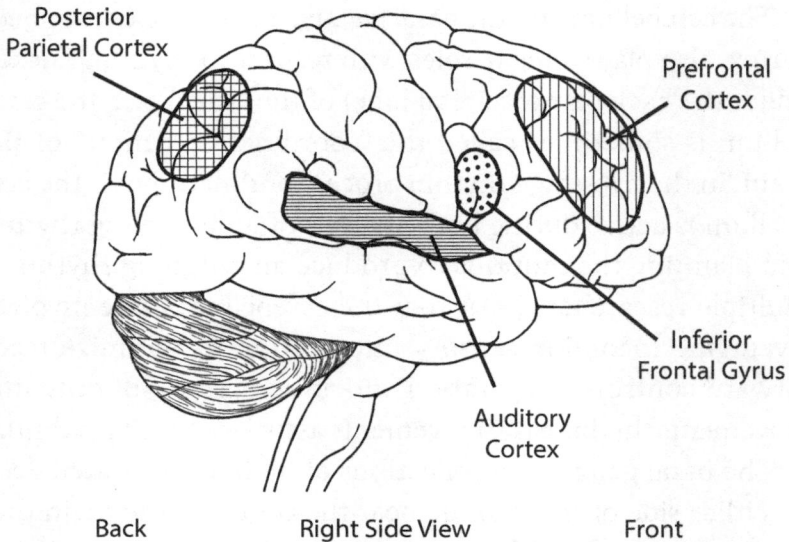

FIGURE 2.6 Regions visible on the right hemisphere of the brain's surface that are active during melody assessment.

Unlike a piece of art resting on an easel, music unfolds over time, and its sounds are ephemeral, as with speech. For this reason, the brain recruits working memory, primarily a frontal (specifically, prefrontal) lobe function.[30] Think of working memory as the ability to keep a string of items in mind so you can grasp the entire sequence. A simple example would be a telephone number, a series of digits. Without working memory, a person would forget the beginning numbers by the time they heard the final numbers. It would be impossible to retain a sequence in one's mind. With music, working memory retains the sequence of musical sounds to make further analysis by the brain possible.

Another key step in deciphering melody is syntax assessment. Without syntax governing which sounds precede or follow which others, sounds would just be random, without any relationship to one another. Syntax enables a series to hang together to be considered as a whole, rather than a hodgepodge of dis-connected parts.

The brain, a pattern-detection machine, searches for patterns among the sounds. It assesses where the patterns comply with and where they violate the rules of syntax established for a particular genre of music or language, both of which are highly rules based.[31] Children readily learn the rules pertaining to the music and language of their culture beginning at a very early age. This is consistent with the dynamic tension between localization and equipotentiality; brain tissue dedicated to syntax exists from birth, and learning is necessary to operate it.[32]

Interestingly, whether processing music or language, the brain makes use of similar real estate when assessing syntax. Its assessment for language syntax primarily takes place in Broca's area, a region of the dominant (left) inferior frontal gyrus.[33] Syntax assessment for music also takes place in the inferior frontal gyrus, but can occur on either side. In music novices, the activity is based in the non-dominant (right) hemisphere. Remarkably, this capacity gradually shifts to the left hemisphere with music training.[34]

Melody assessment is ultimately based on analysis of its pitches, particularly their relationships to one another. The sound wave frequency determines a pitch, also called a tone or musical note. Thus, the brain's analysis of melody involves distinguishing auditory frequencies from one another and then relating them to one another. The capacity to distinguish sound wave frequencies is called **spectral discrimination**.

Only a small fraction of the human population has absolute ("perfect") pitch, the ability to identify single notes in isolation accurately and consistently. So, for the vast majority of people, pitch analysis involves assessing relative differences among successive tones. Specialized sites, clusters of brain cells, are responsible for these pitch evaluations, which take place in the auditory cortex on both sides of the brain. These clusters are distributed throughout the auditory cortex in random fashion.[35] Interestingly, both music and speech activate these clusters, confirming an overlap between music and language assessment in the brain.[36] This functional overlap enables the brain to conserve energy, which is advantageous given the brain's high energy demand.

In addition to being a pattern detector, the brain also functions as a prediction machine by nature. This includes in the realm of melody, where it continuously anticipates what pitch comes next.[37] Clusters of cells specialized for next-pitch prediction have been identified in the bilateral auditory cortex. Similar to relative pitch assessment sites, these clusters are randomly distributed.[38] In contrast to relative pitch assessment sites, next-pitch prediction clusters are selective for music; speech does not activate them.[39] So, analysis of melody is a diffuse function in the association auditory cortex and uses multiple sites—some exclusively for music, some shared with language.

Musical identity is another feature of melody requiring consideration. Melody, more than any other aspect of music, is associated with the identity of a piece of music. Humans recognize and remember music primarily according to its melody. Where does the brain do this? Research points to the right posterior parietal cortex playing a pivotal role in recognition memory for melodies.[40]

In sum, melody is not processed by a single "melody center." In fact, many different regions of the brain participate in melody assessment. Much of the action occurs in the temporal lobes, but melody analysis also activates regions of the frontal and parietal lobes. Some aspects of melody assessment utilize the same brain real estate as language function while others do not; some parts of melody assessment show a right (non-dominant) hemisphere predominance but others do not. Moreover, melody assessment is not static and can change over time. In particular, musical training can influence the process. For example, training can switch which hemisphere of the brain activates for syntax analysis, demonstrating that music can influence brain **plasticity** (see chapter 5).[41] All of these features support the view that music engages the brain expansively.

MANY ACTIVITIES OF THE BRAIN ARE HIDDEN FROM OUR AWARENESS

As shown above, parts of the neocortex are dedicated to the task of assessing rhythm and melody from the time a person is born, and they require learning to function. This is interesting, even fascinating. Yet these details can admittedly seem dry and academic, removed from the enjoyment most people find in music. To understand how the connection of music with sentiment—emotions and feelings—is formed, considering other regions and functions of the brain is required.

Scientists, called natural philosophers until around the 1600s, have for decades analyzed the workings of the brain through keen observation and self-reflection. And why not? Those were the best tools they had at their disposal. Descartes' famous quotation "Cogito ergo sum" (I think, therefore I am) probably best sums up the fruit of their work. In 2012 Erik Kandel, a Nobel Prize-winning neuroscientist, described the fundamental shift that enabled a new perspective on the body's functioning to emerge in the latter half of the nineteenth century. The animating insight of this shift was grasping that much of how the body works remains hidden from view. This recognition subsequently led brain scientists to realize that much of how the brain works is hidden from its own awareness.[42]

As noted, the brain possesses primary sensory (information input) and motor (information output) zones. Primary zones are best thought of as the interface between the brain and the world outside the brain, including the rest of the body. The primary zones organize the world outside the brain in what scientists have termed *maps*. The most famous of these maps are illustrated by Penfield's homunculi (see figure 2.2), arranged in a topographic fashion corresponding to the body's geography.[43] In an analogous fashion, the brain's primary auditory zone represents sounds by way of a tonotopic map, organizing sound waves according to their frequencies (see figure 2.4).

Antonio Damasio, a groundbreaking leader in behavioral neurology, observes that only the brain regions capable of creating maps of the world external to the brain are accessible to consciousness: "The entire fabric of a conscious mind is created from the same cloth—images generated by the brain's mapmaking abilities."[44] Since only brain regions that create such maps are accessible to our awareness, much of the brain is inaccessible to consciousness.[45] This stems from much of the brain not being a primary motor or sensory zone, so not generating maps.

Of note, the word *consciousness* has two meanings, which I label "awake" and "aware." The capacity to be awake long preceded the appearance of human beings and is achieved primarily by structures deep in the brainstem.[46] Consciousness as the full capacity to be aware of oneself and one's place in the world is distinctly human. This quality of awareness is what I generally mean when I use the word *consciousness*. I use the word *unconscious* to mean not awake, and the word *subconscious* to indicate not aware.

Association zones

The non-primary zones of the brain, collectively called association zones, process and analyze the input information that the brain receives at its various primary sensory zones, complemented by information drawn from memory. They formulate output plans that ultimately reach primary motor (movement) zones to prompt action. To reiterate, association zones do not create maps so are not accessible to consciousness.

The simplest association zones, sometimes called secondary zones, are unimodal, further processing information of a single modality received from a primary zone, such as hearing, for example. Multimodal association zones, which are more complex, analyze information from two or more different source types, such as touch (proprioception) plus hearing.[47] Heteromodal association zones, using information received from other zones combined with experience drawn from memory, can make predictions and form judgments. Within the purview of these zones are questions such as Do I need to get outta here because the current situation is a threat? or Is this a feature of music that I value?

The total size of the association zones exceeds the size of the primary zones. The human brain's numerous association zones create the capacity to interpret the data it receives and formulate the plans it makes. These zones operate quickly—because they focus on patterns rather than details—and subconsciously, without people thinking about or being aware of them.

At one of my first lectures in medical school, the chairman of the Department of Pediatrics came to speak to the class. He told us that during our years in medical school and residency, we would spend many hours picking apart and deliberating the fine details of a patient's condition. We would consciously and systematically analyze the evidence we collected through a slow, effortful process. With time and practice, however, we would master the material and then, no longer needing to bring our total attention to it, shift to a pattern recognition method. From that point onward, the diagnosis and treatment plan would seemingly "pop" into our heads, permitting us to move efficiently from one patient encounter to another. This required our thought processes to move from regions of the brain that employ slow methods of analysis based on logical evaluation to regions of the brain that utilize fast methods of analysis based on pattern recognition.

Daniel Kahneman collectively calls the activity of these modes of thinking System 1 and System 2.[48] System 1 thinking focuses on pattern recognition, is essentially always on duty, and works almost effortlessly. But it can arrive at conclusions that are sometimes illogical. That's because the ultimate purpose of System 1 thinking is not rational thought; rather, the value of System 1 thinking is in service to physical needs—detecting threats, seeking nutrition, and in short, helping the body survive and reproduce. These, the chief imperatives of evolution, are System 1's principal tasks.

System 2 thinking—conscious, deliberative thought—is slow and detail oriented, potentially logical, requires effort, and can truly focus on only one thing at a time. System 2 thinking enables humans to unlock the secrets of nature and develop sophisticated technology. It is also the mode of thinking engaged in critical, detached analysis of a piece of music or a work of art. Like System 1, System 2 thinking originated to be of service to the body, say, for organizing a hunting party. Over time, it has developed to the point where the conscious mind can direct it to attain goals and focus on problems beyond biological imperatives. From the perspective of evolution, System 2 represents a truly revolutionary method of thinking that humans possess, yet, despite sitting atop older forms of cognitive processing, it is made of the same brain material that undergirds older modes of thinking, including Kahnemann's System 1. As a result, processes of thought that evolved prior to System 2's appearance affect its functioning. Antonio Damasio explains this elegantly in his masterpiece *Descartes' Error*. Descartes claimed that pure rational thought possesses a unique spirit detached from the body. Damasio, by contrast, wrote, "Nature appears to have built the apparatus of rationality not just on top of the apparatus of biological regulation, but also from it and with it."[49] The failure to grasp this was Descartes' error.

Human thought, no matter how profound or lofty, cannot be divorced from the physical realms that the brain inhabits—the body and the world around it—for two salient reasons: first, rational conscious thought is built upon the brain's maps, which consist of images of the physical world outside the brain; second, rational conscious thought is informed by signals from the body, "gut" feelings, that make it aware of what is important to the subconscious-thinking processes of the brain. These subconscious-thinking processes include genetically determined matters of importance (such as satisfying hunger or avoiding potentially dangerous critters) as well as learned matters of importance (such as which qualities of music a person values). Because of these phenomena, a piece of music touches emotions and stirs feelings even as we simultaneously endeavor to analyze and understand it in a detached and rational manner.

Clarifying hidden workings of the brain

Question: how did neuroscientists learn about the hidden workings of the brain? Answer: slowly, a little bit at a time. Let's begin with the tale of Phineas Gage, one of the most famous patients of brain science. Phineas was a responsible and likable twenty-five-year-old Vermont railroad employee in 1848 when an accidental explosion of dynamite thrust a metal rod completely through the front of his brain. He miraculously survived and returned to work. While he maintained the intellectual and mechanical skills necessary to perform the job, his personality changed markedly. In particular, he no longer acted responsibly and could no longer get along with coworkers. He lost that job and then lost others. He eventually relocated to South America for several years but failed to succeed there either. He ultimately moved to San Francisco—his mother had moved there from Vermont—where he died in 1860.

Nothing remains of his brain. His skull, however, housed in a museum (along with the metal rod!), has been the object of intense study. This investigation suggests that Phineas's orbitofrontal region, the part of the frontal lobe located just above the eyes, was heavily damaged by the blow. Combining Phineas's story with information obtained by examining other patients with lesions at the same location, scientists propose that this area of the brain plays a key role in the ability to make judgments that are called *executive function*. Damasio describes executive function as "[Gage's] ability to plan for the future, to conduct himself according to the social rules he previously had learned, and to decide on the course of action that ultimately would be most advantageous to his survival."[50]

Damasio asserts that these are, by and large, not behaviors and decisions reached after deliberative thought, but primarily real-time assessments made on a "gut" level. What does this mean, on a "gut" level? Areas of the brain that make subconscious executive function judgments act in concert with brain regions that generate emotions, which are plans of action in response to the judgments.[51] Signals from the emotional brain are then sent throughout the body, including to the stomach, via the autonomic nervous system and the bloodstream in order to implement the action plans. The conscious brain, using its map of the body, notices the gut's physical response to the signal. Damasio calls this a "somatic marker."[52] So, listening to our gut, actually means "hearing" a signal from our subconscious brain (inaccessible to our awareness) to the gut (accessible to our awareness because of the brain's map of the body).

What the conscious brain notices in the body—the twisting sensation in our abdomen, the tears in our eyes, the goose bumps down our spine, the tensing of our muscles—prompts what we call feelings: anxiety, sadness, joy, and so on. How does the brain do this? The conscious brain truly experiences, feels, the sensations signaled to it from various parts of the body within its representation (map) of the body. Little does the conscious brain suspect that what it feels is actually the result of activity within the subconscious realms of the brain, about which it has no awareness.

THE SENTIMENTAL SIDE OF MUSIC

How does the above information relate to the musical brain's ability to process and experience the sentimental aspects of music, its emotions and feelings? How, and why, do I, and lots of others, enjoy music? I'll begin with a brief story.

Puccini's "Nessun Dorma" is my favorite male aria. The soloist begins by chanting the words *Nessun dorma* ("None shall sleep") several times in a foreboding manner. Next, he moves to the verse, rendered in a partly sung, partly spoken manner. The intensity of the music then dramatically escalates as the soloist launches into a chorus that conveys soaring passion. After a brief relaxation, the verse and chorus are repeated, although the first half of the chorus is only played instrumentally the second time around. When the vocal artist resumes singing in the latter half of the chorus, the music has already reached its zenith, and the lyrics raise the tension yet another notch. The song ends with the soloist thrice boldly declaring, "*Vincerò!*" (Victory is mine!), followed by the pageantry of an instrumental finish. The element I enjoy most about this aria is being swept up in the moment of transition from the verse to the chorus. I sense goosebumps and feel elated when I hear the shift from the partly spoken verse to the soaring instrumentation and singing of the chorus.

I have not seen *Turandot*, the opera in which "Nessun Dorma" appears, but I have had two memorable experiences with this aria. The first was the one time I saw it performed live. My wife and I attended a concert of diverse classical pieces, a mixture of instrumental and vocal music. "Nessun Dorma" was on the program but, shortly before the concert began, the vocal artist scheduled to appear cancelled due to illness. A substitute was found at the last minute. As he was less known than the originally scheduled performer, no one in the theater knew what to expect. Turned out there was no need to worry. The young man took center stage and, after assuring himself of everyone's full attention, poured forth a beautiful rendition of the aria. He took his bow to great applause and walked off the stage, but the clapping continued until the master of ceremonies brought him back on stage to receive the audience's full and grateful acclaim.

My second experience came from a wholly unexpected quarter. My daughter is a fan of K-pop, Korean pop music. I, on the other hand, had never really appreciated listening to it. Then, one day, my daughter said, "Dad, listen to this." She showed me a YouTube video of a K-pop star named Onew singing "Nessun Dorma." Wow, was it impressive! Not only did the artist solidly hit the notes, he also conveyed the passion of the aria. I thought he did a splendid job. I'm still not a K-pop fan, but I do now acknowledge the talent of K-pop performers.

Forming sentiments

The brain—through a combination of innate (genetic) preferences and acquired (learned) knowledge—assigns personal values to a range of musical features. Based on a person's experiences and tastes, the brain continuously makes predictions about what is about to happen next thanks to its ability to imagine the near future.[53] The association auditory cortex generates predictions and communicates which musical features to reward to a certain brain system. This system is called the reward prediction error system (RPE, or reward system, for short). This process of generating predictions and rewarding the features valued occurs at a subconscious level.[54]

Modern imaging techniques reveal that the striatum, located deep in the limbic brain, is the region activated by the stimulus of hearing music that is valued.[55] Studies also reveal that the striatum and auditory cortex are highly interconnected, consistent with their collaborative roles in this process.[56] In fact, this connectivity is lacking in individuals who do not enjoy music, a condition called musical anhedonia.[57]

Activity in the striatum stimulates release of a group of chemicals, neurotransmitters, especially dopamine, the neurotransmitter most associated with the emotion of approach. Approach means drawing close seeking more of something.[58] In fact, it was observation of rats returning time and again to press a lever to stimulate the striatum in their limbic brains, a pure example of approach, that led to discovering the reward system.

So, what happens when the actual music heard exceeds expectations (predictions) of it? Consider an example: Say a person highly values the sound of a classical guitar. While listening to a recording of a piece of chamber music that follows the typical instrumentation of its genre, a classical guitar unexpectedly begins to play with the chamber group. The actual music has exceeded what the brain predicted. The RPE therefore signals for a suite of neurotransmitters, particularly dopamine, to be released. This initiates the emotion of approach. The neurotransmitters also generate changes in biological regulation, such as the pulse quickens and skin hairs stand up. The person becomes aware of the (up to this point subconscious) reward process by way of the feelings that result: sensing goose bumps, experiencing feelings of elation and joy, and desiring to hear more of it.[59]

Now here's a different example, one of avoidance rather than approach. A person dislikes thriller movies, but a friend twisted his arm to come along and watch a new film of the artform. The dreaded moment of maximum tension arrived à la the stabbing in the shower at the Bates Motel scene in *Psycho*. The unnerving music, full of diminished-seventh chords, served to heighten the frightening effect. His air passages constricted, and he could barely breathe. He felt uncomfortable in his seat. Suddenly, without a moment's reflection, he stood up and walked out of the theater. He relaxed once on the lobby, but wasn't entirely sure how he had gotten there. It must have been an emotion (a program of action directed subconsciously by his limbic brain) of avoidance that came into his conscious awareness by the feelings (discomfort sitting, difficulty breathing) that it generated.

In contrast to approach—the sensation of being drawn in the first example to the classical guitar music—the compulsion to create distance from the thriller movie is called *avoidance*. Also located within the limbic brain, the physical underpinning of avoidance is centered around the amygdala, a structure considered the brain's threat detector. Research has observed activation of the amygdala on both sides of the brain by threatening cues, showing its centrality to avoidance.[60] Consistent with this position, the study of a woman with damage to the amygdala on both sides of her brain showed she was "selectively impaired in the recognition of sad and scary music [whereas] her recognition of happy music was normal."[61] Her general knowledge of music also tested normal, but she characterized "scary" music as peaceful, something normal controls would never do. According to the study authors, she had "an emotion recognition impairment."[62] Other research cited in the same report revealed, "The amygdala can be effectively activated in normal subjects listening to unpleasant music and deactivated by intensely pleasant music." Thus can music stoke, or soothe, the savage beast.

Taken as a whole, these findings indicate that approach and avoidance, though seemingly opposite poles of a single continuum, are actually separate processes served by distinct brain mechanisms.[63] That is, approach results from a set of brain structures and biological regulation mechanisms that can be distinguished from the set of structures and mechanisms that serve avoidance. Activation of the latter set leads to the emotion of avoidance, distancing oneself from the stimulus. This emotion is accompanied by feelings such as fear and dread when the activity of the bodily organs required for avoidance, such as restricted breathing and tensed muscles, comes into conscious awareness.

These contrasting examples—luscious classical guitar music and horror-thriller movies—convey the concepts and mechanisms of the pivotal subconscious processes at work regarding the sentimental side of music. As one listens, the values he or she attaches to musical features become sentiments. Emotional mechanisms move us in response to the music, be it to avoid (run out of the theater) or approach (stand up and call for an encore). Stations in between these extremes are possible and, indeed, far more common. Feelings enter our awareness by way of bioregulation pathways that affect bodily organs. The activity of these organs informs our feelings as it comes into consciousness. These pathways in the limbic and bioregulation layers of the brain are evolutionarily ancient and ever present below the surface of our awareness, even as one strives effortfully and consciously to analyze music dispassionately and objectively in the neocortex. We cannot avoid these sentiments for much of the brain's activity is hidden from our awareness.

MEMORY COMES IN MANY FLAVORS

Now meet Henry Molaison, another famous patient in neuroscience history. Born in 1926, Henry suffered from repeated seizures that no medication could control. It was already known that removal of a section of the brain's temporal lobe can improve seizure control in epileptic patients whose seizures arise in that location. Since Henry's seizures were arising from both of his temporal lobes, he underwent surgery in 1953 to remove a portion of his temporal lobe on both sides. The results were both unexpected and tragic. While Henry's seizure control improved, he could no longer form new memories of events in his life. Henry was trapped in an eternal present.

H.M.—he was known only by his initials until his death in 2008—then became the subject of intensive evaluation over many decades. This testing led to the realization that Henry's episodic memory, his ability to form new memories of events in his life, had been devastated due to the removal of large portions of a structure, the **hippocampus**, from both temporal lobes.

The testing on H.M. also revealed other memory systems in the brain. Most of them are not directly accessible to consciousness, one example being procedural memory, also called muscle memory—that is, the memory of how to execute a complex motor task. Given the assignment to learn how to draw the reverse (mirror) image of an object, Henry was able to accomplish this feat with the same amount of practice that it would take someone with normal episodic memory to learn it, but he could never recall that he had practiced the task.[64] Subsequent testing with other severely amnestic patients, patients with markedly impaired episodic memory, demonstrated similar results with other types of tasks, such as visual recall. Again, the person could learn the new task, but was unable to remember ever practicing it.[65]

Renowned behavioral neurologist Lionel Naccache describes the lack of conscious access to certain types of memory as a dissociation between performance and awareness.[66] As a result, modes of memory inaccessible to consciousness, such as muscle memory, cannot be altered or improved simply by thinking or talking about them;[67] they only change gradually and by repeatedly practicing them with the body. Yoga masters have understood this for generations. One of the reasons for assuming yoga poses is to engage the body: Only in this way can the brain be fully engaged to achieve meaningful change in the ways that it processes information (thoughts) and outputs action (behaviors). True, a degree of learning takes place even when watching somebody demonstrate an action or when imagining oneself performing the action, thanks to brain cells called mirror neurons.[68] This activates only a fraction of the full complement of brain cells required for the entire action, however.[69] That is why, after watching someone demonstrate an action, you then have to practice it again and again to get really good at it.

The bottom line is that the mind—that about which the conscious brain can be aware, or what we recognize as "me"—believes it is in full control of the brain, but much of the brain operates outside the jurisdiction of conscious thought.[70] In reality, the mind's jurisdiction has limits, so changed brain processing and output—that is, changed thinking and behavior—can only be truly achieved through effortful and sustained practice in the physical world.

MUSICAL MEMORY

Another flavor of memory, musical memory, has a distinctive array of qualities. That's because music has the ability to stir up memories, particularly those that are accompanied by strong sentiments.[71] Most people know a song or two that arouses vivid personal memories and emotional responses. Such is the intense emotional connection that musical memories possess.

I didn't care much for college parties—a keg of beer, some cheap eats, and really loud music. Even back then I had trouble understanding people speaking against a noisy background so having conversation with others at a party was always challenging for me. I liked dancing to the music, though, especially with so many great dance songs coming out in the 1960s and '70s.

One of my absolute favorite songs to dance to was "Layla," specifically the up-tempo version released in 1970 by Derek and the Dominos, featuring Eric Clapton.[72] The seven unmistakable notes introducing the song instantly sent the entire room into a buzz. This was the song you most wanted to dance to with that certain someone.

It wasn't just the high-energy music and dancing that made "Layla" special, it was also the shift from all-out energy to a slow dance that made it unique. The transition arrives abruptly, right in the middle of the song, so my partner couldn't say no to dancing slow. And the slow part was quite long, the music creating four dreamy minutes on the dance floor. Every time I hear that version of "Layla," memories of those college days come rushing back to me.

Internally cued vs. externally cued autobiographical memories

How is it that music so effectively brings back such strong emotional memories? A group of researchers at the University of Helsinki reported, "Music can evoke autobiographical memories, which are event specific. Music-evoked autobiographical memories (MEAM) are autobiographical memories specifically elicited when hearing music from one's past and they are typically coupled with the evocation of emotions . . . which are experienced strongly. MEAMs . . . are also likely to be stronger and more [event] specific than consciously recalled [autobiographical] memories."[73]

So, here is a critical distinction between two varieties of autobiographical memory, recalling details about events in one's own life. The first flavor, episodic memory, involves personal memories recalled at will; this is the internally cued pathway. For example, What route did I take when I drove to that restaurant four months ago? or Whom did I meet at the conference last year? Sometimes we have to make an effort, even strain, to recall such details. The second flavor, MEAM, is triggered by something musical from outside the body; this is the externally cued pathway. By chance, you hear a song the way I hear "Layla" and, in an instant, you're transported back to that time in your life. It happens involuntarily and automatically: no conscious effort required, no mental strain whatsoever.

A study of healthy young people showed that a crucial role in the formation and later retrieval of MEAM is played by a region of the limbic brain called the anterior **cingulate gyrus** (ACG).[74] The ACG plays an important role in coupling an event in one's personal life with valence, the importance the brain attaches to the event.[75] Valence can be measured in terms of various factors, including sentiment (such as falling in or out of love), taste (such as I like or don't like that song), motivation (such as how strongly one desires something or someone), or any combination thereof. Music is highly effective at stimulating the ACG to perform this coupling function because music connects strongly with the limbic brain, of which the ACG is a part.

A team of scientists in France summed it up this way: "MEAMs are [event] specific, emotional, fast retrieved [*sic*] . . . and can be considered as involuntary memories. . . . Such involuntary memories are memories of personal events that come to mind spontaneously." Also, "[t]hey pop into awareness without any attempt to retrieve them, these memories are . . . elicited in response to a cue in the environment."[76]

Music that is connected to strong emotions and feelings is more likely to be remembered, encoded in the brain's memory, than music that has little such connection. That's why it's not surprising that most people best remember music from their adolescent and young adult years, times when emotions and feelings, as well as hormones, run high.[77] Music found to be highly satisfying for whatever personal reason is more likely to enter into memory than music one considers bland. With age, some people lose the ability to recall details at will about their lives through voluntary, internally cued efforts, including of songs they once knew. Yet hearing a piece of music familiar to them remains an effective trigger, or external cue, to recall involuntarily not only the memory of the song but also its associated sentiments.[78]

HIGHLY ENGAGING THE BRAIN: CREATING MUSIC

This chapter has covered a lot of ground. The journey began in the neocortex, the newest part of the brain in terms of evolution. This is where much of our thinking occurs, both thinking of which we are aware as well as thinking that occurs subconsciously. The former takes place in the primary zones of the neocortex, regions that make maps of the world external to the brain, including the body outside of the brain. Curiously, the brain has no map of itself, which explains why it is oblivious to the site of perception being the brain itself.

With respect to music, information about sounds arrives in the hearing center in the superior gyrus of both temporal lobes. Each person's primary auditory cortex is organized tonotopically, sensitive to the wavelengths of sounds, and so is a map of the tones it receives. Syntax is then assessed while working memory holds the sequence of tones in mind; these are chiefly frontal lobe functions. Brain regions that assess syntax of tones (music) and syntax of phonemes (language) overlap to a significant degree. This overlap reduces the energy needs of the brain, an organ that demands a tremendous amount of energy to function even during sleep. Melody is primarily evaluated in the temporal lobes, while rhythm engages the motor system in multiple areas of the brain.

Next, judgments about musical elements are made. The auditory association cortex avidly shares this information with the middle brain layer, the limbic brain. As neuroscientist Nina

Kraus puts it, "The limbic [brain] has privileged access to hearing centers."[79] The limbic region is the location of the reward prediction error system, structures critical for deriving pleasure from music. Pleasure is associated with approach behavior. The limbic brain also contains areas responsible for avoidance behavior, chiefly the amygdala. Both approach and avoidance mechanisms link to biological regulation of the body located in the brainstem, the lowest and evolutionarily oldest layer of the brain. This ultimately generates feelings through the brain's awareness of the body responding to bioregulation signals.

In short, music engages nearly the entire brain.[80] This certainly suggests that music benefits the brain, but how can we be certain? After all, maybe music is simply something mirthful. Are there ways to confirm that the power of music upon the brain serves valuable purposes in the physical world? Seeking answers to questions about music's impact using the yardstick of evolution is the goal of the next three chapters.

CONCLUSION

Main points

- The brain is built in successively advanced (or hierarchical) layers, yet the levels from earlier evolutionary stages also continue to function. Ideas and thoughts, feelings and emotions, biological regulation all take place in the human brain. Music stimulates all of these layers.

- Brains originally developed to serve bodies—that is, to respond to challenges related to survival and reproduction. Only eons later could the brain also focus beyond these fundamental concerns.

- Regions of the neocortex associated with musical functions, such as analysis of rhythm and melody, play roles predetermined by genes (nature) while the roles' specific contents are learned (nurture). This is consistent with the dynamic tension between the concept of localization (with each part of the brain destined to perform genetically determined functions) versus the concept of equipotentiality (with education and experience informing what specifics are learned so a broad range of content is possible).

- Music activates the limbic brain, stimulating such emotions as approach and avoidance. These emotions are experienced as feelings, such as enjoyment and fear, through music's connection with the brain's bioregulation functions.

- Music leverages the link between autobiographical events and emotional responses through the process of music-evoked autobiographical memory (MEAM).

- Brains can be trained because they have the capacity to learn even though we lack direct awareness of and access to much of our brain's activity. Such training, a deliberate process, requires a diligent regimen of repetition and practice over an extended period of time.

A few thought-stimulating questions to ponder

- What skills are you willing to practice in order to get good at them?

- Is there a song that stands out to you because its melody or rhythm is so striking?

- Is there a song you enjoy so much that you want to hear it over and over? Alternatively, is there a song that upsets you so much you try to avoid it?

- What pieces of music consistently prompt you to remember events from your past?

PART II

The Evolutionary Impact of Music on the Brain

CHAPTER 3

The Brain's Innate Capacities for Music

"Music is a universal language, and needs not be translated. With it, soul speaks to soul."
—*Berthold Auerbach*

Music benefits the brain. That's a simple statement to make, but how can it be proved? It's not enough to point to the fact that almost everyone enjoys music; more conclusive evidence is needed. One measure of such evidence utilizes the yardstick of evolution. That is, music improves the likelihood of survival and reproduction. Thus, the power of music on the brain is that it promotes life itself.

A major way to establish music's value to the brain from the viewpoint of evolution is to show that the human brain is inherently endowed with capacities to perceive and make music. Innate capacities are encoded in our genes, the currency of evolution, which favors heritable traits that increase an organism's ability to survive and reproduce. Genes that succeed in increasing the chances of survival and reproduction are consistently passed down to future generations; genes that do not increase an organism's ability to survive and reproduce die out or are only randomly passed on.

Beyond being etched into our genes, possessing innate musical capacities implies their evolutionary worth for an additional reason. The human brain requires a tremendous amount of energy to function, even while a person sleeps. Blood vessels continuously deliver 20–25 percent of the body's oxygen and nutrients to the brain to satisfy its energy needs.[1] Maintaining brain tissue is simply too expensive energy-wise to squander on functions that do not contribute real value.

Aaron Copland wrote from his perch as a composer that rhythm and melody come naturally to humans. Brain scientists look for physical brain structures to confirm that humans are born with musical capacities. If such physical structures—brain hardware—can be demonstrated, then it follows that humans are innately music-perceiving and music-making beings.

How is it possible to show that the brain has dedicated structures for music? There are several ways to explore the brain's hardware scientifically. One way is to scan or test people's brains as they engage in musical activities. This maximalist approach, showing activation of multiple brain areas during musical activity, reveals that music avidly activates the brain.

There is a limitation to this approach, however. It does not distinguish between areas of the brain compulsory for a given musical function versus activated brain areas that are not mandatory for that function. In other words, if a brain area is compulsory for a function such as rhythm perception, for example, then the brain cannot perform rhythm perception properly without that area. If the area is not mandatory, however, the brain can get the job done by recruiting other areas and pathways.[2]

A second way to evaluate the brain's hardware is the lesion method, the minimalist approach's gold-standard for identifying the role of a particular area of the brain. If damage to a particular region of the brain results in permanent harm to a given musical capacity, then that capacity is a dedicated function of that particular brain region. This extraordinarily vigorous method can also be applied "in reverse." That is, excessive functioning of an area of the brain can be another way to detect its underlying role. Thus, analyzing overactivity of particular brain regions by meticulous observation of patients can lead to identifying the functions of those regions.[3]

The lesion method has a potential limitation regarding the music-language relationship. Some people hold that music is a byproduct of language. If this is so, a lesion that harms music could actually be a lesion that impairs language, with music decline being a tag-along consequence of the language impairment. This concern may be surmounted by showing that the brain processes music in different locations than it processes language.

THE MUSICAL LIVES OF BABIES

Another avenue altogether to approach the question of whether humans are born with musical brains is to separate nature from nurture. As a practical matter, this strategy can only be applied when studying the very young since they have not yet received any musical nurturing (training). If babies can be shown to possess abilities to perceive music and react to it, then the reasonable conclusion is that these skills are due to nature. In other words, these musical skills derive from the brain's genetically encoded hardware, which, in turn, implies that they provide value to the brain. There are a unique set of challenges to studying babies, but doing so can reveal otherwise unobtainable insights.

The Joy of Little Sounds

A grandchild adds music to our lives.

Our lives changed in the summer of 2022 when my wife and I became grandparents. That August day was particularly poignant for me because my father, his father, and his father's father didn't live long enough to welcome a grandchild into their lives.

Watching our granddaughter respond to music has been one of the joyous wonders of seeing her grow.

We first noticed her recognize our voices and respond by looking straight at us. A little after that, she could be quieted by a lullaby, which made me remember singing lullabies to my children when they were babies. Shortly thereafter, more up-tempo songs began to excite her, and she would happily move her arms and legs. This was especially fun to watch when she took a bath. As she began to vocalize sounds, she could start to "sing along" by saying "pa" or "ma" every so often while someone sang to her. We look forward to seeing her become even more engaged musically as she grows.

Ways to study babies

The study of music in babies is a field that has been very active over the past three to four decades and has produced results demonstrating that infants are naturally attuned to music. Babies can learn from intentional exposure to music as well as from exposure to incidental (background) music.

Consider for a moment some of the challenges researchers face when trying to study babies. For example, how can researchers measure babies' responses to music? Babies obviously can't answer a questionnaire. Moreover, their bodily movements are often so generalized that it's difficult to know what they're actually responding to—as an example, the well-known question "Is the baby smiling at me or just passing gas?" Fortunately, science has a way to deal with this issue, which actually pops up in many study settings, not only those related to babies. The solution is sample size.

If an adult smiles at a baby and the baby smiles back, did the adult's smile elicit the baby's smile? There's no way to know for any single event, but repeating the action many times with numerous adult-baby pairs can determine the relationship between the stimulus (adult smiles at baby) and the response (baby smiles back at adult). Put another way, if the stimulus doesn't cause the response, then the response occurs independently of the stimulus, that is, by chance. For the sake of illustration, think of it like a random coin toss, where each side lands facing up 50 percent of the time. On the other hand, if the latter (baby smiles back at adult) results from the former (adult smiles at baby) then there is a causal link between stimulus and response, and the response occurs more frequently than by chance alone. Again, for the purpose of illustration, let's say the causal response occurs 80 percent of the time.

The magnitude of the difference between the likelihood of the causal event occurring versus the random event occurring—80 percent versus 50 percent in the above illustration—is an important factor in determining the needed sample size. For example, to furnish statistical proof, a causal event that occurs only twice as often as a random event will require a larger sample size than a causal event occurring ten times as often as a random event. So, in determining a sample size, a statistician will ask a researcher how often each event is likely to occur. Since the study has yet to be done, this requires a researcher's educated guestimate. This guesstimate is important because it determines how much time and resources will be required to perform a study. A successful researcher should be commended not only for the results of a study, but also for the skill necessary to manage the time and resources of the study. The statistician deserves commendation, too. A good statistician is not simply someone who analyzes data after the fact; a statistician who can significantly assist researchers construct studies that optimize management of their time and resources is a great asset.

Returning now to studying babies, observing their behaviors is an important way to study them. Beyond smiles, a baby simply paying attention can be a key response to measure as well. The latency, how long it takes for the baby to start paying attention to a stimulus, and the duration of a baby paying attention to a stimulus are important markers of infant behavior. Methods to assess and measure these behavioral responses figure prominently in studies of these youngsters.

In addition to studying behavioral responses, laboratory tests, such as electroencephalography (EEG) to study brain waves, can also be used to observe babies. Brain waves, electrical activity along the surface of the brain, are primarily generated by signaling between brain cells (neurons). This signaling involves the movement of ions—charged particles, such as sodium and chloride, the components of table salt—across cell membranes. Ion movement generates electrical activity, which can be recorded and measured.[4] Electroencephalography is simple, noninvasive, and causes no harm. EEG machines have been used for decades to measure babies' brain waves.[5]

Ultrasound is another frequently used laboratory test. Safe, inexpensive, and easy to use, ultrasound technology has improved so much that it can now display fine anatomic detail.

The anterior fontanelle, the "hole" at the top of babies' heads, allows sound waves to pass into the skull to obtain ultrasound images of the brain. Ultrasound can be used during the first eighteen or so months of life, until the multiple bones of the skull knit together.

In short, the combination of focused observations, straight-forward laboratory techniques, and an ample dose of patience have made it possible to study babies' responses to music. What lessons that have been learned?

Lessons learned about babies

BABIES HEAR BETTER THAN THEY SEE

For starters, the auditory (hearing) system of babies is significantly more developed than their visual system at birth. Research published on this subject dates back at least to the 1960s, with Wolff's demonstration that the earliest infant smile is in response to the mother's voice, not to her face. This is a good point for young parents to keep in mind: talk or sing to your baby because your baby is listening.[6] Talking or singing *to* your baby doesn't simply mean talking *at* your baby. It means speech or song geared to your baby. This is colloquially called "motherese" or "parentese" and scientifically known as infant-directed speech (see chapter 4).

Also noteworthy is that the infant auditory system is already at work before a baby is born.[7] In a clever experiment, a group of expectant mothers late in their third trimester listened to a certain soap opera every day. Observation performed shortly after birth demonstrated that the babies moved significantly more when the soap opera's by-now-familiar theme song played. These infants' heart rates also increased compared with a group of newborns whose mothers had not listened to the soap opera. In this second group of newborns, no increased movement or heart rate was detected when the babies heard what was for them an unfamiliar theme song.

Infants recognize the core features of music—rhythm and a series of pitches (melody)—without any musical training whatsoever. Pitch, the frequency of sound waves, is expressed as musical notes. That babies can detect pitch is unsurprising since, as just noted, they can hear when they are born. What surprises is how fine an ability babies have to detect changes in pitch and the relationship of one pitch to another.[8] This sensitivity speaks to the maturity of the auditory system at birth.

Any single musical note can sound pleasant by itself. It is the relationship of one musical note to another, however, that turns a series of pitches into a melody and some pitch relations are unpleasant to even the youngest human brains. Consonance refers to notes that complement one another to produce a sound pleasing to the ear; dissonance occurs when notes clash and create a harsh, irritating sound.[9] For example, infants prefer to listen to Mozart minuets as Wolfgang Amadeus composed them, rather than to versions of the same pieces altered by substituting dissonant intervals for many of the consonant intervals. The preference for consonance over dissonance has been found even among children born to deaf parents.[10] This strongly supports the position that the ability to perceive music and the preference for consonance are genetically determined (nature) rather than due to environmental influences (nurture).

An interval is the relationship of one musical note to another. Because the human brain focuses on relative pitch, a melody can begin on any note. This is the basis of transposition, shifting a piece of music into a different key. That is, the melody continues using the same intervals but begins on a different starting note. Babies are born with the capacity to identify a melody according to its pitch intervals, so they can recognize a song no matter what key it's played or sung in.[11]

Infants are highly responsive to rhythm, the timing aspects of music encompassing both pattern (meter), which can vary in degrees of complexity, and tempo (speed). At birth, a child has the capacity to learn a wide range of metric complexity. Acclimating to the rhythmic patterns heard in the first year of life, a child develops preferences for familiar ones. Interestingly, if exposed to unfamiliar meters at one year of age, a child still has the ability to gain familiarity with them readily.[12] By contrast, adults are far less able to do this. They become stuck in their habits. The meters they listened to growing up will continue to be the meters they prefer.

Tempo, the speed component of rhythm, is important for the emotional responses it provokes from young children. A slow tempo elicits calm and relaxation, assisting babies to settle down. Caregivers regularly make use of this on a veritable "therapeutic" basis after a baby becomes agitated. This property is also frequently used to help youngsters fall asleep. That lullabies from all cultures are recognized as such by infants supports the idea that this is a global and innate response.[13] Conversely, music with a rapid tempo can stimulate a child's attention, potentially leading to good results (heightened levels of attention or positive affect), but sometimes bad ones (emotional "meltdowns" and feeling overwhelmed). The takeaway here is that tempo modulates the arousal level of babies. Learning how to use this fact optimally is, therefore, a highly useful skill for parenting or caregiving.[14]

It is an irony of music history that arguably the best-known lullaby was composed by a childless bachelor, Johannes Brahms, who set the words of the poem "Wiegenlied" to music in *Lullaby* (1868). He dedicated the song to an old flame on the occasion of her second son's birth.[15] His wonderful musical bequest serves as the template for the lullaby genre with its simple, repeating melodic contour and its lilting waltz rhythm.

Timbre is the unique quality of each sound. Copland called it tone color. Adults listening to music often leverage timbre to identify to which instrument they're listening—"Is that a clarinet or an oboe I hear?" "Would strings or brass more effectively communicate what the composer is trying to express?" "I'm not crazy about the sound of an organ." Infants, however, use timbre to recognize whose voice they are hearing and what emotional message the voice is conveying. The mother's voice, in particular, is vital to recognize, and not only because she is typically the main source of nourishment. She is also a baby's first teacher and the quality of her voice is one of her most important teaching tools. A warm, calm voice communicates a very different message than a cold, agitated voice, and infants respond accordingly to these messages. This is not something that needs to be taught to a child. Children come into this world able to distinguish voices by their timbre and prepared to respond differently to messages as communicated by their tone color.[16]

MUSIC ENGAGES INFANTS MORE THAN SPEECH

Infants are significantly more engaged by music than by speech.[17] Two separate lines of evidence confirm this. First, infants pay more attention to singing than to speech. For example, infants become still and stare at audiovisual recordings of their mother's speech and of her singing. Their attention span, however, lasts considerably longer for her singing than for her speaking.[18] In addition, researchers found maternal singing more effective than maternal speech for relieving infant distress.[19]

Second, babies' movements show that dancing is innate. In fact, the words for "music" and "dance" are the same in many tongues because they are viewed as facets of the same activity. Humans have the capacity to "move with the music"; that is, to coordinate their muscle movements with an external auditory stimulus. Studies of infants show that they engage in significantly more rhythmic movement to musical stimuli—even simple, metrically regular drumbeating—than to speech stimuli.[20] This reveals that humans' capacity to "dance to the music" is innate.[21]

The bottom line is that infants, by nature, prefer song over speech. Their brain hardware is genetically programmed for this. As Robert Zatorre, an internationally renowned music neuroscientists, concluded, "All [the] evidence supports the general idea that the ability to perceive and process music is not some recent add-on to our cognition, but that it has been around long enough to be expressed from the earliest stages of our neural development."[22]

BRAIN DISORDERS REVEAL INNATE CAPACITIES FOR MUSIC

Neurologists, physicians who study and treat disorders of the nervous system, are generally trained to conceptualize the brain from a minimalist perspective. In the case of music and the brain, this means searching for the parts of the brain that are indispensable for music functioning. This outlook leverages the lesion method, taking the position that a lesion—an objective abnormality—is necessary to verify a cause-and-effect relationship between a brain region and a given behavior or skill. Thus, only musical capacities that have been disturbed by verifiable lesions can be considered innate.[23]

Over-expression of music

The lesion method involves investigating the effect(s) of an abnormal part of the brain upon cerebral functioning. The lesion might have been present at birth or may have developed during one's lifetime due to illness or injury. The role, or roles, that a particular region plays in brain functioning can be deduced through observing which functions are reduced when a lesion is present at that spot.

The lesion method can also be applied "in reverse," by investigating the role played by a part of the brain when it is functioning excessively. The most common cause of excessive function of a brain region is when a seizure arises from that location. Such a seizure is called focal, or partial, because only a portion of the brain is involved. A seizure is defined as brain electrical activity that is excessive and firing all at once. It is a brief phenomenon that occurs now and then. When occurring, a focal seizure can provide a window into the functioning of that specific brain area. In the case of a brain region involved with music, a seizure at that site may cause an over expression of its musical function. Examples of music over expression can be divided into two categories: musicogenic seizures, those triggered by hearing music, and musical partial seizures, those experienced as hallucinated music—"hearing" music that is not objectively present.[24]

While any sort of music—classical, folk, pop, even songs sung by a particular singer—can trigger musicogenic seizures,[25] typically only one particular type affects an individual. Interestingly, people who experience musicogenic seizures tend to be musically trained individuals. Brain lesions causing these types of seizures are usually located in the temporal lobe, more often on the right side.[26] As for musical partial seizures, many types of hallucinated sound have been reported by people who have them, but an individual usually experiences one specific type of sound. Some patients "hear" clearly identifiable musical pieces that they can name while others experience only nondescript musical sounds. The most common site for lesions causing musical partial seizures falls in or near the primary auditory cortex, usually on the right side. Contrary to musicogenic seizures, musical partial seizures are not associated with musical training.[27]

Both types of music-related seizures are rare and extremely interesting. Why? They represent examples of seizures generated in a defined (focal) area of the brain (predominantly the right temporal lobe) that are experienced by the individual as increased musical activity. These findings thus supply evidence supporting innate capacity for music located in the temporal lobe, particularly on the right side.

Note that not all seizures cause involuntary movements of the body; only seizures that affect movement areas of the brain do that. Patients susceptible to seizures in brain areas that receive sensory information will experience abnormal sensory activity. The specific symptoms relate to the type of sensory information received in that part of the brain. In the case of music-related seizures, they occur in brain regions that receive information about sounds.

Careful observation of a seizure's signs (what a trained examiner detects) and symptoms (what a patient experiences) has contributed greatly to understanding how different parts of the brain operate. Some of the best-known contributions made in this manner were in the nineteenth-century by John Hughlings Jackson, whose observations of focal seizures enabled him to gain valuable insight into how the brain functions.

The term *epilepsy* is of Greek origin, meaning "to lay hold of," "to seize upon," "to attack." This reflects the ancient view that spirits or gods laying hold of, or seizing, the body caused attacks of epilepsy, what are called seizures in English. These attacks can result in over expression of music function, as described directly above, or in under (impaired) expression of music function when they interfere with the action of the area generating the seizure. Each patient is unique and, sometimes, the condition can sequentially result in over expression and under expression.

Manifestations of epilepsy—during as well as between seizures—reveal a great deal about how the brain works. Specifically, they can help identify the function of the part of the brain they involve, both from their "positive" symptoms and signs (phenomena caused by an excess of function of the part) as well as their "negative" symptoms and signs (phenomena caused by a decrease of function of the part). Indeed, it is important to take both into account for optimal patient care, as the following vignette illustrates.

A neurologist ponders music function

The following is adapted and modified from a fascinating patient case presentation reported in Susan McChesney-Atkins et al., "Amusia after Right Frontal Resection for Epilepsy with Singing Seizures: Case Report and Review of the Literature," Epilepsy & Behavior 4 (2003): 343–47. To bring this medical report "to life" for the reader, I've fictionalized the details of the patient's medical encounters and personal life, created a name for him, and wrapped the story around a conversation I had with an imaginary professional colleague. The interested reader is encouraged to read the endnote referenced at the conclusion of the vignette for it highlights real-life lessons learned from these circumstances.

Some time ago, I attended a professional conference with a group of fellow neurologists. As often happens during the social hour, we began to discuss challenging patients in our practice. My friend Richard related the following tale:

One day, a healthy, right-handed fellow in his early thirties—I'll call him Ben—came into my office. He was married with two young children. He possessed a lifelong interest in music: He played in high school bands, majored in music as an undergraduate, and pursued a degree in music education. He worked as a choir director in the public school system, had a side gig as the director of an adult chorale, and wrote songs when his muses moved him.

In his late twenties, Ben started to have brief episodes of involuntarily singing melodies completely unfamiliar to him. The episodes occurred without warning and only every once in a while. The episodes didn't bother him; in fact, he rather enjoyed them because some of the melodies were clever and he could use them as the basis for composing songs. He thought of the episodes as his subconscious creative self finding a means of expression.

Several years passed and then, during one of these episodes, Ben couldn't stop slapping his thigh with his left hand. He rationalized that he was just keeping time with the music that had come into his head. When it happened a second time, he again ignored it. When it happened a third time, he began to worry.

When I saw Ben in my office, I tried to reassure him, but I was concerned about the possibility of seizures. I scheduled him for a couple of tests. His MRI scan showed a lesion in the posterior [rear] portion of his right frontal lobe. The EEG revealed seizure activity potentials in the same region.

When he returned with his wife for the follow-up appointment, I said, "Ben, your EEG, which shows the electrical activity of the brain, indicates that these episodes are seizures coming from the same part of the brain where the spot on your MRI scan is. The seizures can be controlled by medication, so I'll send a prescription to your pharmacy. And I'm concerned about what that spot on your MRI is. It needs further looking into."

"What could the spot be?" Ben's wife naturally asked.

"I don't know at this point," I responded, "but without further testing I can't rule out the possibility that it's some sort of a small tumor."

The word *tumor* upset both Ben and his wife, of course—no patient or spouse wants to hear that word—so I took the time to reassure them that a tumor doesn't necessarily mean a bad cancer. "We'll repeat the brain scan in six months to see if the spot is stable or changing," I told them.

When six months rolled around, Ben was ready for the repeat scan. Although the lesion on the MRI was stable, he asked to have it removed. He consulted a neurosurgeon who sent him for an extensive battery of tests that included confirming the precise location responsible for his seizures, defining the areas of his brain that controlled strength on his left side, and ensuring that none of his language faculties would be disturbed by removing that part of the brain.

When the school year ended, the day for Ben's surgery arrived. The lesion was expertly removed. He sailed through the surgery and was home in just a few days. Over the ensuing months of his summer vacation, his seizures stopped and his medication was gradually phased out.

By the time the summer break was over, Ben felt fine and returned to his choral directing. He quickly ran into a problem though. He was no longer able to sing notes correctly. He could hear them properly, but couldn't produce them accurately. Then he noticed having the same issue with rhythm: He could perceive others doing it correctly, but he couldn't. Naturally, he was deeply concerned when he came to see me in the middle of October.

I said to him, "Ben, I'm glad to hear you're not having any more seizures. And it's good you had the surgery. Study of the tissue removed from your brain showed a nest of cells that didn't end up in the correct place when your brain formed before you were born. These cells have stayed quiet for many years, but now show signs of becoming active and could one day turn into a cancer."

"That's good to know," he responded, "but I'm having a problem at work, and I can't write songs anymore either."

He proceeded to describe the difficulties he was encountering with singing melodies and executing rhythms properly. I didn't know what to say other than, "Be patient. Your brain probably needs more time to heal."

The months passed, but his problem didn't get any better. The chorale's annual holiday concert had to be canceled. Then the arts and music director of his school district told Ben that he couldn't stay on as the choir director. He applied for disability and, fortunately, was awarded a monthly stipend to make up for some of his lost salary. He stayed home and slipped into depression, dissatisfied with the results of the surgery even though he was now seizure-free and a potential cancer in his brain had been removed.

I said to Richard, "Sometimes you can't win for losing. You diagnosed his problem precisely, stopped his seizures, and prevented him from one day having a brain cancer. What more could you have done?"

Richard took a deep breath and sighed. "I've been rerunning the entire scenario in my head and I've become rather philosophical about it. I'm pleased how the doctors, myself included, did what we were trained to do. But we didn't fully consider the possible consequences from Ben's perspective because we failed to consider whether his musical livelihood might require the brain tissue we removed. Even though he was seizure-free and no longer had a potential cancer in his brain, he was dissatisfied with the results of the surgery on account of not being able to work as a choir or chorale director afterward. He lost the ability to earn a living doing what he truly loved in life—sharing music with others.[28]

Lessons learned from the minimalist approach

The minimalist approach emphasizes learning about the brain by studying the abnormal functioning that results from an identifiable brain lesion. In the case in the preceding vignette, Ben's music over expression brought him to medical attention and a diagnosis of seizures. He developed music under expression following his operation, losing the ability to perform certain musical functions after removal of the area of his brain that had caused the over expression. In medical speak, he was left with an **amusia** (decrease or loss of musical function) for expressing pitch (melody) and rhythm after surgery; he retained his **receptive** pitch and rhythm skills. Ben needed the area of his brain that was removed for pitch and rhythm expression to work properly. This indicates that these functions are innate brain capacities performed in that specific area of the brain. His receptive capacities to perceive pitch and rhythm were not disturbed, indicating that they are functions based elsewhere in the brain.[29]

AN ESSENTIAL BRAIN NETWORK FOR MUSIC

Scans show that music activates numerous, widely spaced regions of the brain. The maximalist approach does not make a distinction between regions that are obligatory for music function versus those that activate but are optional for music function.

Where, then, does music function occur in the brain according to the minimalist perspective? The answer requires asking the question: Where in the brain are lesions found in people who have permanent amusia (music dysfunction)? The majority of studies done in pursuit of answers to this question involve patients who have had strokes. Thanks to neuroimaging, strokes can be readily located in the brain. Interestingly, many of the deficits initially caused by strokes—the symptoms patients report and the signs health care providers observe at the stroke's onset—resolve on their own, generally within a few months and sometimes even within a few days or weeks.[30] This is why final determination of a patient's deficits after a stroke is typically deferred until three months after the stroke occurred.[31]

A group of Finnish researchers examined patients with amusia caused by stroke. Soon after stroke onset, they identified lesions causing amusia in numerous areas on both sides of the brain. They then waited for three months before conducting a follow-up examination and found that most patients' music deficits had cleared up during that time. For those patients whose amusia did not resolve, imaging at three months located their strokes in areas of the right hemisphere that closely mirror the language network of the left hemisphere. The researchers concluded that the affected areas of the right hemisphere contain the brain structures indispensable for music function (see appendix A).[32] Thus the lesion method demonstrates that these specific structures have innate capacity for music.

What about those other areas on brain scans that "light up" or activate from music? Strokes signify irreversible damage, yet strokes in those areas cause short-term musical problems but not permanent musical deficits. Why is that? This remains fertile ground for ongoing study and dialogue between maximalists and minimalists to generate more questions and test hypotheses. For example, perhaps these areas perform operations necessary for music perception or production, but the brain has redundancy and can therefore ultimately do without them while still delivering the product by accessing alternate routes; this is akin to taking a detour to get around a road that's blocked. Another option is that the brain has sufficient plasticity to develop new pathways for these necessary functions during the three months following the stroke. Yet a different possibility to explain why amusia may clear up on its own is that the area of damage, while permanent, decreases in size so much during those three months that no deficits persist; think of the way scars shrink over time. Also, given that stroke patients with music plus language deficits are often excluded from amusia studies, perhaps additional stroke locations that cause amusia are underrepresented in this analysis because they also cause **aphasia**, abnormality of the brain's language functioning?[33] The bottom line is that more work needs to be done. Neither camp is "right" while the other is "wrong," and the dialogue generated among scientists with differing viewpoints advances the knowledge gained in the fascinating adventure of understanding the brain.

Tone deafness

Tone deafness is a disturbance of melody.[34] It is a lifelong deficit of pitch discrimination, or in other words, an inability to distinguish two different but closely related musical notes from one another. As a result, one's ability to perceive or produce small pitch changes is impaired.[35]

The term *deafness* is misleading. By definition, the diagnosis of tone deafness requires a person's hearing to test in the normal range. Tone deafness is not a problem with one's hearing at all. In fact, the connections from the outside world to the inner ear and from the inner ear to the brain's auditory cortex work normally in tone deafness. So, to find its cause requires looking more deeply into the brain.

The term *tone* in this context indicates that the condition relates to the brain's ability to distinguish musical notes correctly—specifically, telling one note apart from its neighboring note. *Semitone* is the term for the interval between one note and the note immediately preceding or following it on the chromatic scale, the series containing all twelve semitones of an octave.[36] Individuals with normal pitch perception can readily distinguish the interval of a semitone. For example, when the note A and then the note just above it, A# (A sharp), are played, a person with normal pitch perception can readily tell the difference between the two. In fact, normal infants can do this as well.[37] By contrast, individuals with tone deafness require a difference of at least two full semitones to detect a pitch change. In the case here, that would be the auditory distance between the note A and the note B above it. This is a gap large enough to interfere with the brain's perception of subtle pitch change, meaning that it can impair one's ability to detect or produce the series of notes that form a melody.

IMAGING TONE DEAFNESS

Tone deafness is a **congenital** condition, meaning people are born with it. It stems from dysfunction of the right arcuate fasciculus (figure 3.1), one of the brain's white matter pathways, the hardwires that permit communication among the various grey matter modules of the brain. On the left side of the brain, the arcuate fasciculus belongs to the language network and connects Wernicke's area to Broca's area, modules for speech comprehension and production, respectively. On the right side of the brain, the arcuate fasciculus belongs to the innate network dedicated (and therefore indispensable) to music, as found by the Finnish researchers mentioned above.

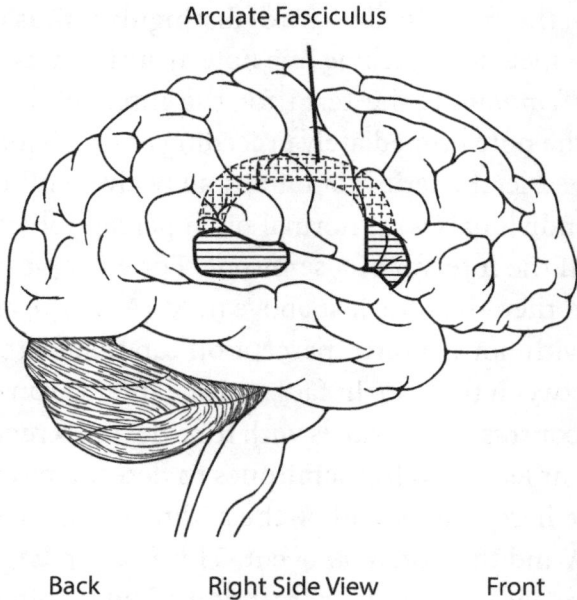

FIGURE 3.1 Location of the right arcuate fasciculus.

Showing that tone deafness correlates with dysfunction of the right arcuate fasciculus utilizes **tractography**, a brain imaging technique that produces scans revealing the presence of activity in white matter pathways. White matter pathways are called tracts. The scans are examples of *functional* imaging— demonstrating the presence of physiologic activity—not images of anatomy. Tractography imaging has shown that individuals with tone deafness have impaired functional connectivity— absent or markedly reduced communication—along the right arcuate fasciculus.

The upper diagram in figure 3.2 represents intact functional activity—that is, normal communication—along both the left and right arcuate fasciculus. This is in contrast to the situation in the lower diagram, where activity along the left arcuate fasciculus is intact, but is absent along the right arcuate fasciculus. The absence of physiologic activity along the right arcuate fasciculus in the lower cartoon does not necessarily mean that the arcuate fasciculus is missing; the structure may be present, but not working.[38] This imaging study links absent or markedly reduced functioning of an identifiable brain structure, the right arcuate fasciculus, with a dysfunction of music, tone deafness. It therefore illustrates that tone deafness correlates with impaired functioning of the right arcuate fasciculus.

Summing up

To recap the above sections, brain disorders confirm that the brain has innate capacities for music. This is demonstrated through meticulous observation of the symptoms and signs of these conditions together with insights provided by modern neuroimaging. This powerful combination reveals brain regions

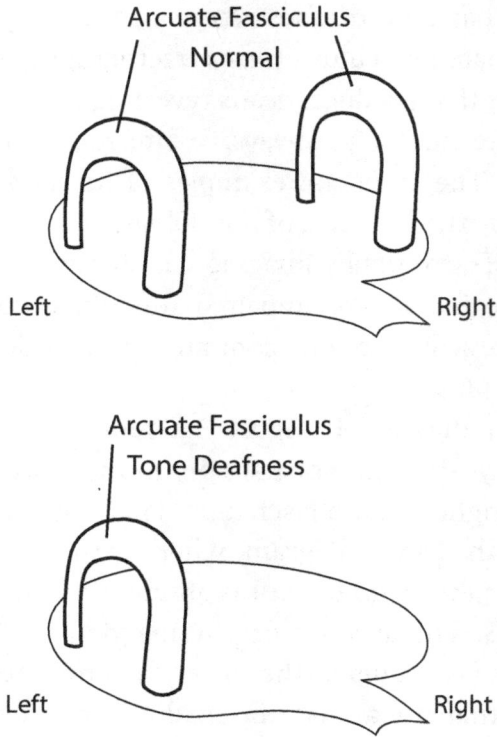

FIGURE 3.2 Representation of a functioning (*top*) and impaired (*bottom*) right arcuate fasciculus. The arcuate fasciculus is an important white matter pathway. This functional image is an example of tractography, assessing the presence of activity in a white matter pathway (as opposed to showing the pathway's anatomy). Adapted from Psyche Loui, Catherine Wan, and Gottfried Schlaug, "Neurological Bases of Musical Disorders and Their Implications for Stroke Recovery," *Acoustics Today* 6, no. 3 (2010): fig. 2, https://doi.org/10.1121 /1.3488666, with the kind permission of Psyche Loui.

that are indispensable for music function. The fact that disorders of the physical brain can cause permanent impairment of music function reveals that the brain has dedicated real estate for music, physical underpinnings that reinforce the concept that the brain has innate capacities for music.

THE MUSIC–LANGUAGE RELATIONSHIP

What is the relationship between music and language in the brain? If music is a by-product of language, as some assert, then music results from the same brain areas that serve language. If, however, music is entirely distinct from language, then different brain areas serve music and language. Is either one of these polar-opposite positions correct?

A brain scan primer

Tackling the relationship between music and language requires some familiarity with displays of brain imaging, such as CT (computed tomography), MR (magnetic resonance), and PET (positron emission tomography) scans.

Suppose there's a magic knife that can make a perfectly straight cut through the entire brain in any of the three planes of space. Just as magically, the cut pieces can then be precisely reassembled. With the pieces separated, the newly exposed surfaces can be examined. That's what someone looks at when studying a slice, or a section, of a brain scan. A vertical cut from front to back, to make left-side and right-side slices, creates what is called a sagittal section; a cut from left to right, to form a front slice and a back slice, creates a coronal section; and a horizontal cut to make a top slice and a bottom slice creates an axial section. The axial, or horizontal, plane is the most useful one for this discussion.

An axial series, scans made by a succession of horizontal cuts through the brain, creates a stack of slices from the top of the head to the base of the skull. Each slice is displayed with the front of the brain at the top of the picture and the back of the brain at the bottom of the picture. As the middle image from the stack is generally the most informative, a template of this image can be made, showing the parts of the brain in diagram form (figure 3.3). The words *left* and *right* at the bottom indicate the brain's respective hemispheres. The labels of the four lobes of the neocortex apply to both sides of the brain. Of note, the temporal lobes are drawn proportionately larger than their actual size to enhance the information value of the template because much of the activity related to music occurs there.

A template such as this serves several functions. It can be used to diagram the results of an individual scan. It is also

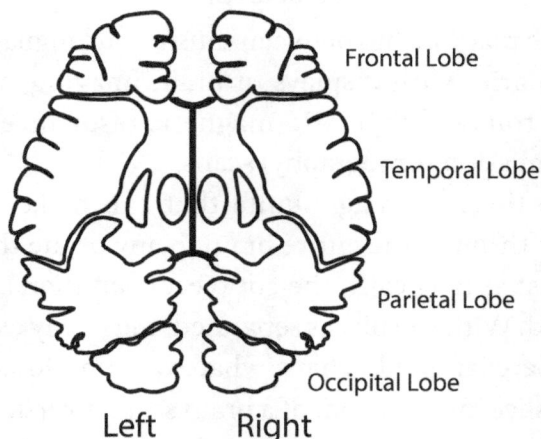

Frontal Lobe

Temporal Lobe

Parietal Lobe

Occipital Lobe

Left   Right

FIGURE 3.3   Template of a brain axial image. Brain structures close to the center of the figure but not labeled on the template: the corpus callosum (the major communication route between the two hemispheres), indicated by the dark pillar shape along the midline (the imaginary line dividing the brain into left and right hemispheres); the thalamus (a major relay station for sensory and motor information), shown as ovals on either side of the midline; and the basal ganglia (important for motor control), the banana-shaped forms on either side of the midline.

The cerebellum, located under the occipital lobes, is not shown in this template.

Adapted from Lauren Stewart et al., "Music and the Brain: Disorders of Musical Listening," *Brain* 129, no. 10 (2006): fig. 3, https://doi.org /10.1093/brain/awl171, with the kind permission of Timothy D. Griffiths.

useful for overlaying the findings of multiple scans—such as the scans of many different people—to create a single, composite scan of a group.

Brain structures close to the center (midline) of the figure but not labeled on the template: the **corpus callosum** (the major communication route between the two hemispheres), indicated by the dark at the center of the figure; the thalamus (a major relay station for sensory and motor information), shown as ovals on either side of the midline; and basal ganglia (important for motor control), the banana-shaped forms on either side of the midline. The thalamus and basal ganglia are gray matter structures (collections of brain cell nuclei); the corpus collosum is a white matter pathway (wires that connect gray matter structures to one another).

Sihvonen looks at strokes and disorders of music

Studying the music–language relationship begins here by looking at **acquired** amusias. The term *acquired* means the problem was not present at birth. In fact, the function was normal until an illness or injury impaired it. The majority of acquired amusias occur as a result of strokes. Other, less common, causes include (but are not limited to) brain tumors and head trauma.

Aleksi Sihvonen, a scientist at the University of Helsinki, Finland, studies brain mechanisms of music and language.[39] His research team performed scans on patients who suffered strokes. Each of the three figures that follow represents a composite scan of the patients in the group. This is accomplished by consolidating and superimposing the scans of the patients in each group to form a single image. The striping—horizontal in the case of music, vertical in the case of language—identifies areas of the brain which have permanently impaired function due to damage caused by strokes.

Figure 3.4 shows where strokes are located in patients who have amusia as compared to those without amusia. This comparison is done in order to isolate specifically where strokes causing amusia are located within this sample of patients. These strokes only affect the right hemisphere. The researchers excluded patients having any evidence of **aphasia** from this group.

Figure 3.5 shows where strokes are located in patients with aphasia as compared to those without aphasia. This comparison is done in order to isolate specifically where strokes causing amusia are located within this sample of patients. These strokes only affect the left hemisphere. The researchers excluded patients having any evidence of amusia from this group.

Figure 3.6 compares the above two groups of patients: those with amusia only versus those with aphasia only. The compari-

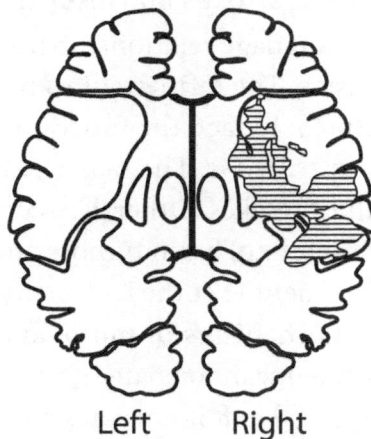

Left   Right

FIGURE 3.4 Location of stroke damage to the brain (striped area) in a group of amusics, people with abnormal music function (excludes aphasics, people with abnormal language function). Adapted from Aleksi Sihvonen et al., "Neural Basis of Acquired Amusia and Its Recovery after Stroke," *Journal of Neuroscience* 36, no. 34 (2016): 8876, https://doi.org/10.1523/JNEUROSCI.0709-16.2016, with Aleksi Sihvonen's kind permission.

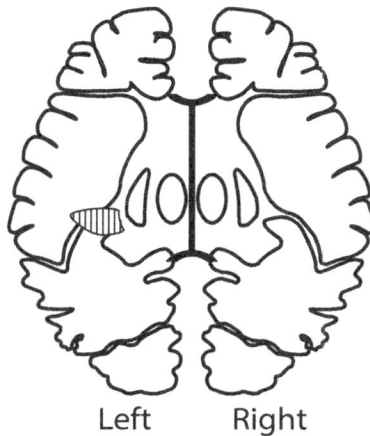

Left   Right

FIGURE 3.5   Location of stroke damage to the brains (striped area) of a group of aphasics, people with abnormal language function (excludes amusics, people with abnormal music function).
Adapted from Aleksi Sihvonen et al., "Neural Basis of Acquired Amusia and Its Recovery after Stroke," *Journal of Neuroscience* 36, no. 34 (2016): 8876, https://doi.org/10.1523/JNEUROSCI.0709-16.2016, with Aleksi Sihvonen's kind permission.

son reveals that individuals with a "pure" amusia ("pure" in this instance means no component of aphasia) have a lesion of the right hemisphere whereas individuals with a "pure" aphasia (meaning no component of amusia) have a lesion of the left hemisphere. This supports what may be called the conventional wisdom that music function occurs in the right hemisphere while language function occurs in the left hemisphere.

Griffiths looks at strokes and disorders of music

Some researchers have adopted meta-analysis—an approach consolidating the results of multiple studies—to look at the concept of relating disorders of music function to stroke locations. A meta-analysis can be a powerful tool, especially since it

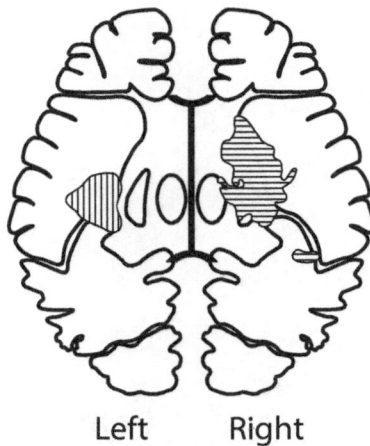

Left   Right

FIGURE 3.6   Location of stroke damage (horizontal striping) to the brains of a group of amusics, people with abnormal music function, versus stroke damage (vertical striping) to the brains of a group of aphasics, people with abnormal language function. Author's note: The asymmetry of the stroke volumes—the composite size of strokes causing amusia is noticeably larger than the composite size of strokes causing aphasia—may be due to the relatively small total sample size (seventy-seven patients). A larger sample size may have resulted in greater symmetry of the volumes of the stroke territories.
Adapted from Aleksi Sihvonen et al., "Neural Basis of Acquired Amusia and Its Recovery after Stroke," *Journal of Neuroscience* 36, no. 34 (2016): 8876, https://doi.org/10.1523/JNEUROSCI.0709-16.2016, with Aleksi Sihvonen's kind permission.

provides a much greater sample size than a single study can provide.[40]

Tim Griffiths, a professor of cognitive neurology at Newcastle University, England, and his research team reviewed data from dozens of patient reports.[41] The reports described the kinds of abnormal music perception these patients were experiencing along with information about where in the brain their strokes were located. Only perception of music was studied;

musical expression, such as singing or playing an instrument, was not assessed in this analysis. To facilitate interpretation of such a large amount of information, the data from all of the study participants were superimposed into single, composite illustrations plotted onto the template introduced in figure 3.3.

The templates in figure 3.7 focus on the two core elements of music—melody and rhythm. The findings on each template represent brain locations where at least half of the studies submitted for meta-analysis showed a lesion for that musical category. For individuals who had a disturbance of melody, each circle filled with horizontal striping on the left template indicates a site where at least 50 percent of the brain scans showed a stroke; for individuals who had a disturbance of rhythm, each circle filled with horizontal striping on the right template indicates a

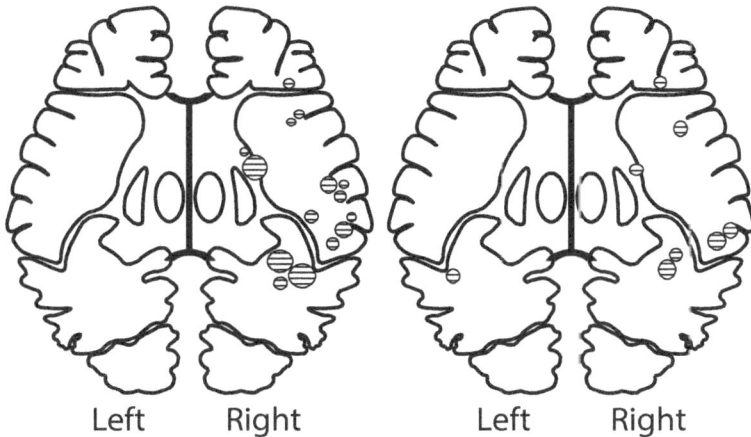

Left   Right   Left   Right

FIGURE 3.7   Location of strokes (striped circles) causing melody (*left template*) and rhythm (*right template*) disturbances. The researchers caution, "[These diagrams] should not be equated with maps of the whole network required for a particular function; rather, they represent the critical components of normal networks." Adapted from Lauren Stewart et al., "Music and the Brain: Disorders of Musical Listening," *Brain* 129, no. 10 (2006): fig. 3, https://doi.org /10.1093/brain/awl171, with the kind permission of Timothy D. Griffiths.

site where at least 50 percent of the brain scans showed a stroke.

All in all, Griffith's approach yielded results consistent with Sihvonen's approach. In Griffith's meta-analysis, strokes that cause amusia (music impairment) were overwhelmingly located in the right hemisphere. With but one small rhythm exception, on the right template, the circles are all located in the right hemisphere. Given that strokes causing aphasia, language impairment, lateralize to the left hemisphere (as in Figure 3.5), the lesion pattern for it is different from the lesion pattern causing amusia.

In short, the results of Sihvonen and Griffith's studies reveal that music is served by different parts of the brain than those that serve language. Scientifically speaking, this is evidence of the dissociation of the neuroanatomical correlates of music and language. This fancy term means that a person can lose language function without losing music function and vice versa because language and music functions occur in different parts of the brain.

An example of losing language function without losing music function involves a famous Russian composer, Vissarion Shebalin, who suffered a stroke in 1953 and again in 1959. His strokes impaired much of his language abilities.[42] On the other hand, his musical skills remained undiminished and, just a few months before his death from a third stroke in 1963, he completed his fifth symphony. This symphony was described by Dmitri Shostakovich, an even more famous Russian composer, as "a brilliant creative work, filled with highest emotions, optimistic, and full of life."[43]

Another telling example of losing language function without losing music function was Louis Victor Leborgne, the patient whose aphasia was described by the famed Parisian neurologist Pierre Paul Broca.[44] While his aphasia limited his vocabulary to a single word—"tan"—Monsieur Leborgne reportedly could still sing *La Marseillaise*, the French national anthem.

Brown examines music and language

It's true that some strokes cause only music dysfunction, and some strokes cause only language dysfunction. It's also true that some strokes cause mixed—music and language—dysfunction. This suggests a degree of overlap between the music and language operations in the brain. To understand this requires digging deeper and shifting focus from the brains of individuals with strokes to the brains of normal individuals.

The lead author of this research study is Steven Brown, who studies the neural basis of the arts at McMaster University in Hamilton, Ontario, Canada.[45] Over the past several decades, brain scientists have learned that the amount of blood flowing to an area of the brain increases rapidly when that part of the brain becomes active; conversely, when a part of the brain takes a rest, blood flow to it decreases rapidly.[46] This means that measuring blood flow to a brain region reveals whether that part of the brain is active or resting at a given point in time. The images obtained by Brown's research team used a positron emission tomography (PET) scanner to measure cerebral blood flow. This technique calculated the amount of blood flowing to a region of the brain as participants in the study performed tasks assigned by Brown and his team.

Figure 3.8 shows brain areas active during a melody generation task. For this particular task, participants heard five to six seconds of a novel melody, that is, a melody they had never heard before. They then had to improvise the next five to six seconds of the melody on the spot. Each participant was scanned during the improvisation phase in order to assess what areas of the brain received increased blood flow (indicating increased activity) during the task of melody generation.

Note that "increased" is a relative term. That is, saying blood flow to a region of the brain is increased requires specifying increased compared to what. The phrase "melody generation–rest" in the caption of Figure 3.8 means that the task of melody generation is being compared to the task of resting (specifically, "resting with eyes closed," according to the authors' report). The individual scans of all of the study participants were then combined into one consolidated group scan. So, this template represents a composite of the group members' scans showing the areas of the brain activated by melody generation—as measured by increased cerebral blood flow—but inactive at rest.

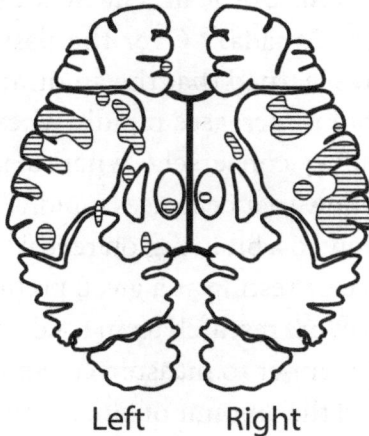

Left   Right

FIGURE 3.8 Melody generation—Rest: A composite scan showing areas of the brain activated by melody generation, as measured by increased cerebral blood flow, but inactive at rest.
Adapted from Steven Brown, Michael Martinez, and Lawrence Parsons, "Music and Language Side by Side in the Brain: A PET Study of the Generation of Melodies and Sentences," *European Journal of Neuroscience* 23, no. 10 (2006): 2797, https://doi.org/10.1111/j.1460-9568.2006.04785.x, with the kind permission of Steven Brown.

This template shows that music, as gauged by the spontaneous generation of a melody, recruits (activates) areas on both sides of the brain. Not only that, a gestalt overview suggests that the two hemispheres are rather similarly active, the right hemisphere perhaps to a slightly greater extent. Perhaps the studies of Sihvonen and Griffiths discussed above gave the impression that music belongs exclusively to the right hemisphere, while language is solely the province of the left hemisphere? Ah, the brain is full of surprises.

The activity shown in figure 3.8 overwhelmingly occurred in the middle tier of the brain, not much in the front (at the top of the image) or back (at the bottom of the image) tiers of the brain.[47] Sihvonen's and Griffith's scans showed this pattern as well (figures 3.6 and 3.7, respectively). As a general rule, the visual system occupies the back (posterior) tier of the brain, and cognitive skills—such as reasoning, deliberation, and forming judgments—take place in the front (anterior) tier of the brain. In Brown's study, all the participants performed tasks with their eyes closed, so little activity in the posterior tier was expected. They generated melodies exclusively by vocalizing the sound *da*, sung at different pitches, so little in the way of judgment was required. The melodies had to be produced spontaneously so deliberation and reasoning were not required.

Figure 3.9, captioned "sentence generation–rest," shows brain areas active during a sentence-generation task, but inactive at rest. Participants listened to a novel sentence fragment that lasted about five to six seconds and then, on the spot, had to generate a conclusion to the sentence of similar duration. The scanning was performed in a manner similar to the previous template.

What are the findings when a sentence generation task is compared to the brain at rest? As with melody generation, the

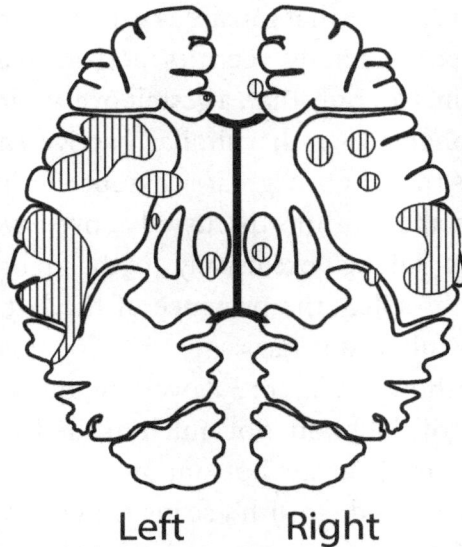

Left   Right

FIGURE 3.9 Sentence generation–Rest: A composite scan showing areas of the brain active during a sentence-generation task, but inactive at rest.

Adapted from Steven Brown, Michael Martinez, and Lawrence Parsons, "Music and Language Side by Side in the Brain: A PET Study of the Generation of Melodies and Sentences," *European Journal of Neuroscience* 23, no. 10 (2006): 2797, https://doi.org/10.1111/j.1460-9568.2006.04785.x, with the kind permission of Steven Brown.

sentence generation task activated areas on both sides of the brain. A gestalt overview suggests that the two hemispheres were recruited to a similar extent, the left hemisphere perhaps slightly favored this time. Once again, the activity overwhelmingly took place in the middle tier of the brain. Might the small region of activity in the frontal region represent the brain's faculty of judgment monitoring the participants' choice of words, to ensure that no "naughty" ones slip through?

It would really be interesting to look at an overlay of these two templates to learn which brain regions were activated solely by music (as assessed by melody generation), which were activated only by language (as assessed by sentence generation), and which were activated by both music and language. For that, see figure 3.10.

What does this overlay show? It looks like areas activated solely by melody generation are bilateral but somewhat more numerous on the right side while areas activated solely by sentence generation are bilateral but somewhat more numerous on the left side. Strikingly, there are plenty of areas on both sides of the brain that are activated by both music and language,

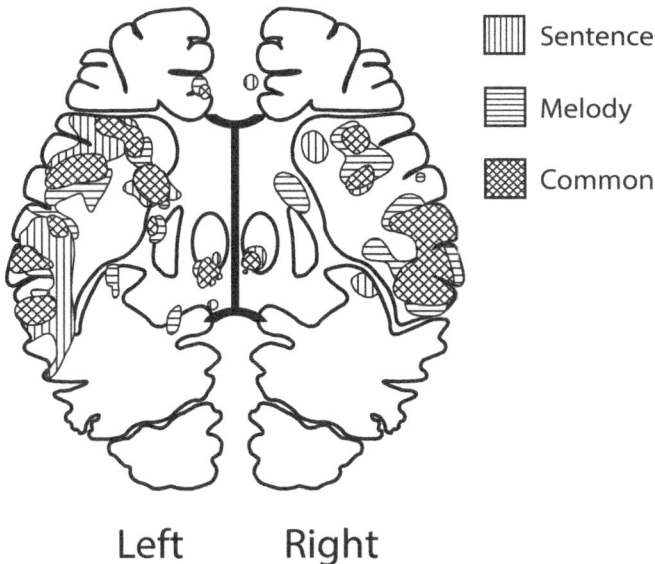

Left   Right

FIGURE 3.10   Melody generation and sentence generation: An overlay of two templates showing brain regions activated solely by music (assessed by melody generation), areas activated only by language (assessed by sentence generation), and areas activated by both music and language.

Adapted from Steven Brown, Michael Martinez, and Lawrence Parsons, "Music and Language Side by Side in the Brain: A PET Study of the Generation of Melodies and Sentences," *European Journal of Neuroscience* 23, no. 10 (2006): 2797, https://doi.org/10.1111/j.1460-9568.2006.04785.x, with the kind permission of Steven Brown.

revealing that multiple brain regions serve both music and language. What explains this? The brain shares resources for these mutual functions to maximize its energy efficiency. This is called shared processing. For example, on the "input" (receptive) side, both music and language require analysis of sound signals arriving to the brain from the ears,[48] both require keeping track of a sequence of sounds, and both are based on a set of rules, syntax—called grammar in the case of language—that governs how sounds (pitches for music, phonemes for language) are to be combined. On the "output" (expressive) side, both require planning and control of complex movements, such as voice production and articulation, involving the mouth, tongue, throat, breathing apparatus, and so on.

Then there are areas that are unique to music and areas that are unique to language. The brain has parallel—rather than shared—processing for functions that are unique to one or the other. This is particularly the case for semantics, the extraction and insertion of meaning, as Brown and colleagues make clear in their report.[49] Semantics for music is predominantly the domain of the right hemisphere whereas semantics for language is predominantly the domain of the left hemisphere.

To recap the observations of Brown's group, music function and language function activate both hemispheres of the brain. Numerous aspects of the brain's role in the perception and production of sounds are similar for music and language, so are handled by the same brain regions. This is termed shared processing. Shared processing with respect to music and language takes place on both sides of the brain. On the other hand, when elements of music or language require specialized treatment, the music and language elements are handled by different brain structures through parallel processing. As regards music and language, parallel processing takes place for the most part on

Meaning in Music

The concept of meaning with respect to language is familiar—language is all about the communication of meaning—but what about "meaning" as regards music? Patel, in *Music, Language, and the Brain*, chapter 6, presents an excellent discussion of this question. He begins by citing a number of possible definitions of meaning in music and settles on the following: meaning exists when perception of a musical event brings something to mind other than the musical event itself (p. 304). He then explores several possible examples.

First, musical elements bring subsequent musical elements to mind. This is called *anticipation* and is a concept alluded to in chapter 1 (see pp. 34–35, where it is termed *prediction*). The complexity of the anticipation increases as one acquires greater knowledge of and experience with music. **Second, musical elements express moods.** While auditory vibrations "out there" in the environment possess certain objective features, mood is ultimately human interpretation "in here" in the brain, so it is influenced by culture and personal experiences. Nonetheless, some musical elements that evoke the same (or similar) moods cross-culturally have been identified, such as joy being associated with fast tempo and low melodic complexity. **Third, music evokes sentiments.** One of the major reasons people listen to music is its impact on their emotions and feelings. Research shows that music not only evokes everyday sentiments, it also evokes emotions and feelings that can be measured physiologically in the form of chills. Music is the most effective art form for evoking chills, although other art forms, such as painting, can also do it. **Fourth, music spurs motion.** Humans are capable of synchronizing bodily movement to a musical rhythm (see chapter 4). This property, known as entrainment, is linked to the vestibular system's ability to perceive motion. Stimulation of the vestibular system induces a subjective sense of movement even if the body remains objectively still. **Fifth, music cultivates social associations.** I refer to this as the pro-social component of entrainment. Some consider this property to be the most meaningful aspect of musical experience altogether. Patel also points out that advertisers leverage this attribute to sell merchandise: for example, customers buy more expensive wine when classical music is playing in a wine shop. **Sixth is music's capacity to imitate sounds found in nature,** be they of the environment or of animals. This is called *sound painting.* Among the best-known examples of such pieces are Beethoven's Sixth Symphony (Pastorale) and works by Debussy such as "Claire de Lune" and "La Mer." The above are Patel's major points, although not exhaustive of his discussion. We may consider them to be neuroscientific perspectives of musical meaning since Patel is a neuroscientist.

Music took on wholly different concepts of meaning during the communist era in the Soviet Union. Reflecting on the impact of Shostakovich's music during the communist era, a violinist from Leningrad said, "What we were really thinking, we couldn't talk about. But when you listened, Shostakovich's music explained everything that had happened, in our hearts and in our minds." Similarly, the poetess Anna Akhmatova wrote, "Only music speaks to me when others turn away their eyes," in her poem "Music," which she dedicated "To Dmitri Dmitriyevich Shostakovich, in whose epoch I live on earth."[50]

opposite sides of the brain—music-specific elements in the right hemisphere, language-specific in the left hemisphere.

Meaning in language and music

How do semantics—information extraction and insertion—differ between the two hemispheres, and what accounts for these two systems being on opposite sides of the brain? Start with the fact that hearing has unique properties among the senses. It can operate at a distance (you can hear something that's far away), function at night, and even go around corners. It's also quick. "You hear anywhere from 20 to 100 times faster than you see," wrote neuroscientist Seth Horowitz, "because the brain's auditory circuitry is less widely distributed [i.e., is more compact] than the visual system and sound signals don't have as far to travel in the brain."[51] This was already the case among non-human primates equipped for generally listening to sounds. The additional demands placed on human auditory function— to process music and language—magnifies the importance of this compactness in the human brain.

Language is a form of communication based chiefly on the temporal discrimination of sounds, how rapidly one phoneme can be distinguished from the next. Music is a form of communication based primarily on spectral (sound wave frequency) discrimination of sounds, distinguishing the pitch of one sound from the next.[52] The temporal discrimination skills of the brain enable humans to distinguish many distinct language sounds (phonemes) per second. Think, for example, of radio ad announcers who speak astoundingly fast, the ones whose rapid-fire voices spew out the disclaimers at the end of commercials that often end with "void where prohibited." Because transfer of information between the two hemispheres takes time— approximately twenty-five milliseconds for each one-way trip— temporal discrimination has to be lateralized to keep the network for language spatially compact. If the left and right hemispheres were to co-manage temporal discrimination processing, the back-and-forth communication between the two sides would take up so much time that we would miss a great deal of what was said. Instead, one hemisphere developed superior temporal discrimination to handle the assignment. And what happened in the other hemisphere? Superior spectral discrimination developed.

Another reason for hemispheric specialization is the uncertainty principle. In physics, the Heisenberg uncertainty principle holds that the more we know about a moving object's position, the less we know about its velocity and vice versa. This is because we cannot simultaneously have full knowledge about the position and speed of a particle.[53] The uncertainty principle is thus a trade-off between two complementary variables, two variables about which we cannot have fully accurate knowledge simultaneously. In this instance, the variables are position and velocity.

In acoustics, the Joos uncertainty principle states that greater temporal discrimination of a sound comes at the expense of spectral discrimination and vice versa.[54] As with the Heisenberg uncertainty principle, the Joos uncertainty principle derives from the limited precision that comes when measuring two complementary variables, in this instance frequency and time.[55] This explains why parallel processing, rather than shared processing, developed to handle these dissimilar tasks. Furthermore, to reap the full benefit that flows from keeping each of these functions of the auditory system spatially compact, these skills developed in functionally specialized centers positioned close to the auditory cortex on opposite sides of the brain.

Functional specialization requires specialized anatomy

The separation of temporal and spectral discrimination is an example of parallel processing: one brain region attends to one function while a different brain region occupies itself with the other. Moreover, temporal and spectral discrimination are located in opposite hemispheres—the dominant hemisphere for temporal processing; the non-dominant hemisphere for spectral processing—so the brainscape for each function can be as compact as possible.

Figure 3.11 illustrates the concept that the brain must physically separate its regions for temporal and spectral discrimina-

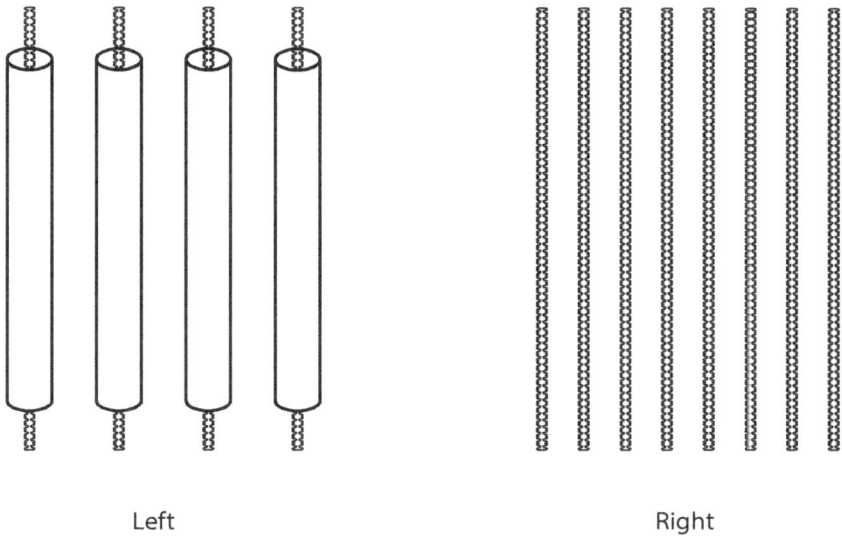

Left   Right

FIGURE 3.11   Differing brain anatomy for temporal (*left*) and spectral (*right*) discrimination. Myelinated axons (represented by straws around pipe cleaners) allow for faster signal transmission which favors temporal discrimination. Unmyelinated axons (pipe cleaners without straws) can be more tightly packed, permitting a greater range of frequencies to be recognized favoring spectral discrimination.
Adapted from Robert Zatorre, "Neural Specializations for Tonal Processing," in *The Cognitive Neuroscience of Music*, ed. Isabelle Peretz and Robert Zatorre (Oxford University Press, 2003), 239, with the kind acknowledgment of Robert Zatorre.

tion. It is based upon the work of Robert Zatorre, from McGill University, who explains that different brain anatomy is needed to perform these different functions. The diagram shows the contrast between the axons (represented as pipe cleaners) of the auditory regions in the brain's two hemispheres. Compared with the right side, the left hemisphere's axons are more thickly insulated with myelin (represented by straws surrounding the pipe cleaners). Like insulation surrounding a copper wire, the myelin allows for faster nerve signal transmission speed. This results in superior temporal discrimination, vital for comprehending language. By contrast, the right hemisphere's auditory cortex axons are poorly myelinated. Therefore, they can be more densely packed. This allows for greater spectral discrimination—fundamental to understanding pitch—since each nerve is tuned to a unique frequency. This occurs, however, at slower nerve signal transmission speed because the axons lack myelin. Zatorre notes, "These structural features would lead to relatively fast conduction but poor spectral resolution on the left, and to poorer temporal resolution but greater sensitivity to fine spectral differences on the right."[56]

In sum, music and language share a number of aspects of brain processing. To optimize its energy efficiency, the brain utilizes similar regions for these operations whether it hears music or language; this is called shared processing. On the other hand, to process elements unique to music or elements unique to language, the brain makes use of parallel processing—discrete, specialized areas of the brain—to perform those tasks. In the case of music and language, those separate areas are on opposite sides of the brain to preserve the compactness of the auditory system. Such brain asymmetries—the different skill sets of the two hemispheres—developed as the brain's "solution to the need to optimize processing . . . in both temporal and frequency domains."[57]

The examples and studies discussed over the course of this chapter present evidence that there are specific regions of brain hardware dedicated to music and that essential music functions are genetically hardwired in the brain. Sources of this evidence span the entire age spectrum from infants to the elderly (most strokes affect elderly people) include congenital and acquired brain conditions, take into account normal subjects as well as individuals with brain lesions, consider brain under expression and over expression, and investigate key similarities and differences between music and language processing. Taken as a whole, this body of evidence supports the proposition that the human brain has innate capacities for music.

But of what value are these abilities to the demands of evolution? How is it that musical brains better enable humans to survive and reproduce? Seeking answers to these questions, the next chapter will explore music's power to provide evolutionary benefit.

CONCLUSION

Main points

- The human brain has genetically determined innate capacities for music. This implies that the brain values music, especially since maintaining brain tissue is too expensive energy-wise to squander on unnecessary functions.

- Hearing is significantly more developed than vision at birth, and a baby's earliest smiles are in response to its mother's voice.

- Both the maximalist and the minimalist approaches to brain research are valid and vital; the dialogue generated by their differences advances the scientific understanding of the brain.

- The lesion method—the gold standard of the minimalist perspective for identifying the function of a specific brain area—can be applied when the activity of that area is reduced (under-expression) as well as when it is excessive (over-expression)

- Several major lines of inquiry demonstrate that music is a dedicated function of the human brain, rather than a by-product of language:

  ○ Babies are substantially more engaged by music than by speech. This preference is present in newborns, requires no training whatsoever, and appears before language skills emerge.

  ○ A person can lose language function without losing music function and vice versa. This demonstrates that the brain possesses areas dedicated to music that differ from those dedicated to language. In fact, key components of music and language operate in opposite hemispheres of the brain. A telling example of this is the language network of the left hemisphere and the music network of the parallel area of the right hemisphere.

  ○ Amusias not associated with aphasia lateralize to the nondominant (right) hemisphere. The nondominant hemisphere is also specialized for spectral analysis of sound, a critical component for music processing.

- Music and language are based on brain capacities that arose in response to the physics of sound. Music and language coevolved in service to different human needs.

A few thought-stimulating questions to ponder

- What differences have you observed between the musical tendencies of young children versus those of adults?

- If you have raised children, did you sing to them when they were young? What did you like to sing to them?

- If music did not exist, how do you imagine your life would be different? Explain your points of view.

- Do you know anyone who possesses superior musical ability? What do you think accounts for it?

CHAPTER 4

Music's Evolutionary Benefit to Humans

I love to hear a choir. I love the humanity, to see the faces of real people devoting themselves to a piece of music. I like the teamwork. It makes me feel optimistic about the human race when I see them cooperating like that.
—*Paul McCartney*

Music's impact upon the brain confers evolutionary benefit to human beings. Evolutionary benefit improves the chances of the species to survive, through natural selection, and to reproduce, through sexual selection. Music accords evolutionary benefits to humans in a host of ways. A vast subject, these ways are explored in this chapter as four themes, or hypotheses: communication, caregiving, entrainment, and procreation. An essential fact to keep in mind while reading about these hypotheses is that music is universal; it exists in all human societies and all peoples of the world sing and dance.

THE COMMUNICATION HYPOTHESIS

Communication is generally viewed in terms of the exchange of information between two or more parties in, or very nearly in, real time. In this context, language is a means of communication that no one doubts confers evolutionary benefit. That's because sharing information via language increases the chances of survival and procreation. Admittedly, music cannot compete with language in terms of exchanging bits of novel information clearly and efficiently. Music's message is typically inexact, whereas meaning in language can be highly precise.[1] Yet, there are some settings where music can express information more effectively than language.

The power of combining synchronized voices

Singing can communicate to greater numbers of individuals and over greater distances than speech. (Comparison with music is limited here to spoken language without amplification by electronics. Written language, a recent development in human history, does not engage the auditory [hearing] system, so is excluded from this comparison.) It's a matter of scale. Only one person can speak and be understood at a time. Everyone has experienced the chaos that ensues when many people try to speak simultaneously. Conversely, multiple people can sing together, and the scope of their reach can be additive. This relates to the timing constraint inherent in music. When people sing in time with one another, their voices are amplified. Think of a choir: The voice of any one member may be barely audible, but everyone singing together can be awe inspiring. Thus, the power of music through song to deliver a message can exceed the power of language through speech.

This can be represented graphically in a straightforward manner. Two sound waves traveling in the same medium, air for this discussion, will interact with one another. **Interference** is the technical term for this interaction. If the waves are in phase (i.e., in sync) with each other, as when people sing in unison, their amplitudes add together. The interference becomes "constructive" and can rock a building (figure 4.1).

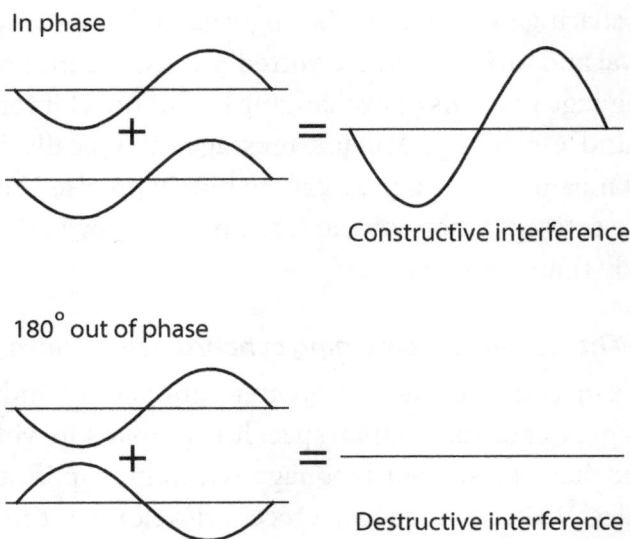

FIGURE 4.1 Interference patterns representing effectiveness of combining sounds. Effectiveness, which can be thought of primarily as clarity in combination with volume, is measured by the distance from the center horizontal ("zero") line in either direction. The flatline indicates sounds that are out of sync so do not combine effectively.

On the other hand, if the sound waves are out of phase—as when people sing out of sync with one another—their amplitudes subtract from each another, yielding disarray or "destructive" interference. This is similar to when people try to speak over one another. This chaos is sometimes a desired, usually comical effect in an opera or musical when, for example, multiple performers each relate their unique point of view of a situation simultaneously through their own song even though no one is listening to—or can understand—what any of the others are saying.

For people to sing together effectively, they rehearse many times to stay in phase, and the content of the information must be highly fixed and adhered to, so all of the participants know exactly what to sing. This lacks the flexibility and spontaneity that language affords. In other words, music is not as good as language for communicating novel or alternative information. That said, music can be more effective in conveying other types of information.

Music for accumulated and shared knowledge

Music can be a very effective method, even superior to language, for transmitting or reinforcing accumulated knowledge. Dan Levitin, who became a music neuroscientist after a career in the recording industry, refers to such music as "songs of knowledge."[2] Simply stated, prior to the appearance of writing and literacy, passing information from one generation to the next was far more effective through song than prose—primarily because people remember lyrics better than they recall prose. This can be readily proven in a couple of easy ways: First, recite as many poems as you can recollect from high school; then, sing as many songs as you can recall from that same time period. It's not even close, right? You can recall many more songs than poems. Second, have a friend or family member choose a song. Next, speak the lyrics without singing them. Stop after you hesitate even for a second. Now, sing the lyrics, and again, stop after you hesitate. How did you do: Did you recall the lyrics better by singing them than speaking them?

Levitin wrote, "The very poor typical recall of text stands in stark contrast to the very good typical recall of song lyrics"[3] Why is this? It's because the human brain is poor at remembering things word for word, but a song sets up multiple simultaneous structures that guide the brain to recall the lyrics. Foremost is rhythm, the timing component of music, dictating the number of syllables in a verse and also stress placement. Melody, the primary source of its identity, gives a song direction and emphasis regarding which words carry greater weight than others. Lyrics leverage poetic devices such as alliteration, metaphor, and especially rhyme. Rhyming imposes major constraints on end-of-verse and sometimes mid-verse word choices. Together, these "mutually reinforcing constraints" create a well-defined structure for a song such that memorizing each and every word becomes unnecessary.[4] This is more effective and efficient for the brain.

More, and varied, brain regions are activated in remembering lyrics than prose. The distinguished music therapist Concetta Tomaino wrote, "Brain science tells us that when words are embedded in music, they are processed in a way that is quite different from the way we process ordinary speech."[5] For example, music engages motor areas of the brain related to the component of timing. Interestingly, Alzheimer's disease barely affects motor areas of the brain until its very advanced stages. The ability to access the brain by activating motor areas—through singing, dance, or even simple movement—offers a way for demented individuals who otherwise show little verbal interaction to become animated participants by moving to a familiar song.

In preliterate societies, songs were a better way to communicate information because people remember lyrics better than prose. This remains true of children in the modern world. Children learn their ABCs to music. They learn to count by using music—such as counting the bottles of beer on the wall. They learn about the parts of a bus, the parts of their body, the history and traditions of their culture through music. Levitin also reminds, "Don't forget that the psalms, as originally composed, were set to music."[6]

The quote from Levitin three paragraphs above—"The very poor typical recall of text stands in stark contrast to the very good typical recall of song lyrics"—continues with "especially in the case of long epic ballads "[7] Levitin's use of the expression *long epic ballads* can be interpreted as referring to fairly lengthy songs that describe historical events in the form of a continuous narrative that minimizes the use of repeating refrains.[8] This genre was an important method of sharing knowledge before a widely literate public could read newspapers and magazines. By means of epic ballads, traveling troubadours could deliver information that a few members of a community memorized. These members then recited it again and again to spread the knowledge throughout the community.

Gordon Lightfoot's remarkable ballad

Songs like "The Wreck of the Edmund Fitzgerald" (1976), written by Gordon Lightfoot, are rarely created anymore. In the spirit of a traveling troubadour, Lightfoot penned this superb example of the ballad genre just a month or so after the events it describes occurred. He sang the song often as he toured, effectively spreading the story of the shipwreck. His biographer Nicholas Jennings noted, "[Lightfoot's] name is synonymous with timeless songs about trains and shipwrecks, rivers and highways, lovers and loneliness."[9]

"The Wreck of the Edmund Fitzgerald" is based upon the events of November 10, 1975, when the freighter SS *Edmund Fitzgerald* sank in a storm on Lake Superior, claiming the lives of all twenty-nine aboard. Lightfoot begins the song with a powerful declaration about the lake's fearsome might, which has been recognized ever since the region was long ago populated by peoples of the First Nations.[10]

Moving to the particular, Lightfoot introduces the ship, its industrial mission transporting iron ore, and its capable crew, set against the backdrop of the eerie dangers of November on

Lake Superior. Having set the scene, he creates a sense of the crew's gradual awareness of the changing weather. Lightfoot conveys imagery of the mounting wind and rain, using sharp, descriptive adjectives like *slashing* and *freezing*. Next the cook speaks, at first to announce that the sea is too rough to serve dinner and then later that evening to bid his fellow crew members farewell. The cook's sense of impending doom kicks the listener right in the gut. Finally, the ship loses power and sinks.

At this point, Lightfoot could describe the agony of the crew members' death, but instead he switches gears: he lowers his voice and becomes religious or philosophical. He asks where is the Almighty's love in such a circumstance? He emphasizes the human tragedy that has befallen the bereaved family members. He returns to the theme of Lake Superior's formidable strength and compares it with the other Great Lakes.

In the final stanza, Lightfoot describes a memorial service for the crew members at the "maritime sailors' cathedral."[11] Sailing being such a dangerous line of work, churches serving mariners were common in port cities. Importantly, the manner in which Lightfoot sings the entire piece—at a measured pace in waltz meter using a chanting style without any refrains—gives the impression of listening to church bells tolling on the first beat of each measure throughout the song. This method produces a dramatic impact.

Lightfoot concludes the song by returning to where he began: proclaiming the awesome might of Lake Superior with words close to those with which he started, the unmistakable message being that the power of man is no match for the power of nature.

All living things must confront Mother Nature's fickle whims, a sudden storm on Lake Superior being one of them. Earth is a highly dynamic ecosystem with an environment constantly in flux. To survive, species have to meet the challenge of adapting to persistent change.

The original biological basis of such adaptation was genetic mutation. A member of a species that happens to possess a genetic trait to facilitate adaptation to changing conditions makes it better positioned to survive and pass this trait on to future generations. Possessing such a new or mutated gene is a random event. If no members of a species possess such a gene, the species may not survive a changed environment. In fact, many species have gone extinct by this route.

Over the course of evolution, learning developed as a second method for adapting to change. Learning greatly accelerates the process of adaptation because it bypasses the need to wait for a necessary, but random, genetic mutation. The human species has excelled at learning to the point of being able to adapt to diverse and changing environments around the world. From the perspective of evolution, learning has enabled humans to be immensely successful.

Passing learned information from one generation to the next requires a deliberate process of teaching. This is where songs of knowledge play a major role. Through "The Wreck of the Edmund Fitzgerald," Lightfoot imparts historical material and instills knowledge in a readily remembered manner. As a result, the information can be easily recalled and passed on so another member of the group or the next generation can learn it and benefit from it. In this way, Lightfoot's song demonstrates how music can be the optimal means for communicating information, thus showing how music confers evolutionary benefit to our species.

Charles Aznavour's immortal ballad

"La Bohème" (1965), a wonderful song of knowledge by Charles Aznavour, is a ballad that has achieved mythic status in the French-speaking world.[12] While "The Wreck of the Edmund Fitzgerald" recounts the events of a single day, "La Bohème" relates the essence of a time—the bohemian era in Montmartre, one of the most storied neighborhoods of Paris.

Bohemian-era Montmartre became known as a stomping ground for artists in the late nineteenth and early twentieth centuries. Many of them were drawn there by the cheap rents. While perhaps longing for fame, only a few, such as Pierre-Auguste Renoir and Pablo Picasso, attained it. It was those who never tasted success, the long-forgotten anonymous bohemians, who provided the bottom-up energy of the district. Montmartre was the stuff of their dreams, and the artistry of "La Bohème" gives voice to them. The song features the reminiscences of an inconnu, an unknown no-name, who arrives in bohemian Montmartre full of spirit and overflowing ambition, only to have them stymied by the realities of trying to earn a living as an artist.[13]

What was bohemianism (la bohème)? Bohemians saw themselves as distinct from the broader society and its turmoil. Some viewed bohemianism as a temporary station in life, something they would outgrow. Others considered themselves the vanguard of a future world culture and wanted to live nowhere but "Bohemia."[14] Think hippies in the 1960s. Moreover, although formally incorporated into Paris in 1860,[15] Montmartre remained a proudly unconventional hilltop village intent on preserving its independent identity. This provided a perfect perch from which to observe and confront the encroachment of urban industrial society. Think San Francisco in the 1960s.

A key facet of Montmartre's popularity was actually its broad appeal to the general public. Through the enormous creativity of its residents, Montmartre offered commentary and reflection on society at large without throwing too many stones at it. It attracted a variety of visitors from Paris and beyond, including Jean-Martin Charcot, the most famous neurologist in all of nineteenth-century Paris who enjoyed frequenting Montmartre for its entertainment. Charcot introduced a young Sigmund Freud to the neighborhood's night spots.

Impressionism, the famous genre that Montmartre nurtured, was initially mocked by the staid art establishment, but would go on to conquer the world. Of all Montmartre's well-known artists, Henri de Toulouse-Lautrec most famously captured the neighborhood's ambience through paintings and prints of the entertainers who worked there. He leveraged novel, modern technology: particularly lithography, which offered easy mass production and expanded the capacity to advertise the district's entertainment venues. Photography also contributed mightily to Montmartre's fame as images of it spread rapidly around the world. Music, too, underwent creative changes, ranging from the impressionist music of Claude Debussy to the novel sounds created by Erik Satie. And in dance, the can-can became virtually synonymous with Montmartre.

Presenting himself as a starving young artist in bohemian Montmartre, Aznavour begins the song by describing its late nineteenth-century setting. Perched on the highest hill in the region, Montmartre retained the feel of an airy village full of flowers and gardens, a stark contrast to the stench of overcrowded Paris below. He goes on to relate that he was unable to earn a living despite sleepless nights toiling at his easel to perfect his technique. As a result, the young artist and his girlfriend depend on the money she can earn by posing nude as a model. Later, when a cafe owner takes pity on him and barters a hot meal for a painting, the young artist celebrates with his friends. Despite winter's bracing chill, he and his comrades are in high spirits living their bohemian lives.

In the final stanza, Aznavour's lyrics wax philosophical. Decades have passed and the once young man has moved on. During a chance visit to his old haunts, he observes that the neighborhood is no longer what it used to be and that it's impossible to return to those days of being poor as a pauper yet wondrously happy. Not only can he never go back, there is no back to go to because time has but one direction—forward. The key message of the song is to explore each decade of life fully, absorbing its lessons, and to accept the stage of life one is in since the past no longer exists.

Yet Aznavour, in the last line of the song, offers a puzzling irony. In hindsight, he claims that it meant nothing at all to have spent his twenties—crazy yet formative years for so many— living in bohemian Montmartre. How can that be? A person who has lived life fully, absorbed the message of each decade, and who accepts the present, has truly gained wisdom. This apparent contradiction is a key ingredient of what makes "La Bohème" so captivating, for the moment after Aznavour finishes singing the final words, the music's mood abruptly shifts from forlorn retrospection to whirling excitement. In public performances, Aznavour would lose himself, entranced in dance, celebrating his grasp of the enigma. He senses the eternal glory of bohemian Montmartre—its cultivation of creative youthful potential—deep in his soul. He accepts that he, as a maturing individual, naturally aged out of the neighborhood. The idea of bohemian Montmartre never disappears, however. It awaits the next generation to discover it and renew it.

This is a generous legacy to leave to future generations. By relating a compelling story in song, Aznavour goes beyond knowledge to impart wisdom. He communicates important lessons about life that improve a person's chances to make it in this world. "La Bohème" demonstrates that such powerful ideas are best communicated through music. Its survival value is why natural selection favors music.

THE CAREGIVING HYPOTHESIS

Bearing and raising children stands at the confluence of natural selection and sexual selection. According to the caregiving hypothesis, music facilitates caring for offspring by parents and other adults, making survival of the children more likely. As previously noted, infants are ready for music from birth (see chapter 1). In fact, the maturation of musical skills in very early childhood is, in and of itself, a point in support of the caregiving hypothesis. Turning the argument around, what does the hypothesis say about the perspective of caregivers? All over the world, caregivers in all human societies use music when attending to babies. In short, "music is ubiquitous in caregiving."[16] Caregivers use music to soothe babies and to arouse them. Caregivers are also ready to modify their voice after their infants are born and even have a special, musical way of vocalizing to babies that transcends specific languages.

Technically called **infant directed speech**, but commonly referred to as "motherese" (and now also "parentese" as fathers speak it too), this manner of vocalization is characterized by warm timbres, slow tempos, high pitch levels, and expressive dynamics (variations of volume). To be clear, parentese is a technique for delivering grammatically correct language to an infant; it is not baby talk in the sense of uttering nonsense syllables like goo-goo-ga-ga or calling objects by silly names like shoozie woozies.

Parentese has been studied among many language groups, including tonal languages such as Chinese. A study of eight Mandarin Chinese–speaking mothers showed that they use parentese with their offspring, just like mothers who speak nontonal languages. A study from the University of Washington observed, "[The] pattern of results for Mandarin Motherese is similar to that reported in other languages and suggests that Motherese may exhibit universal prosodic features."[17] **Prosody** refers to voice stresses and inflections that impart sentimental features to spoken language.

The voice is not the only attribute that parents naturally modify when in caregiving roles. A study of deaf mothers in Japan showed that they modify their signing when interacting with their babies. The author wrote, "When communicating with their infants, the [deaf] mothers used signs at a significantly slower tempo than when communicating with their [adult] friends. They tended to repeat the same sign frequently, and the movements associated with each sign were somewhat exaggerated. Thus, a phenomenon quite analogous to Motherese in maternal speech was identified."[18]

Infant directed speech (IDS) benefits caregivers and infants alike. Researchers in Paris noted that as a package, "The prosodic patterns of IDS are more informative [to infants] than those of ADS [adult directed speech] and they provide infants with reliable cues about a speaker's communicative intent."[19] This means that caregivers communicate their intentions to infants by way of prosody—sometimes called the musical qualities of speech—in IDS. This is superior to using ADS, the way adults speak with and listen to one another, as ADS overwhelmingly relies on the meaning of words (language semantics) to communicate intention.

The observation that IDS is found throughout the world strongly supports its benefit to and for infants. A 2013 meta-analysis revealed that IDS promotes infants' affect, attention, and eventual language learning.[20] It also promotes emotional bonding between caregiver and child. While the meta-analysis acknowledged that the cognitive benefits of IDS have been more widely studied than its affective (emotional) aspects, it also observed that IDS prosody reflects caregivers' emotional content and meets infants' preferences for discerning meaning. This means that caregiving has developed according to what works best for and with infants.

The infant's preferences reinforce the optimal techniques for the caregiver to use. IDS is thus part of an interactive loop between infant and caregiver. Indeed, infants remember and gaze longer at individuals who address them using IDS rather than ADS. In addition, IDS likely plays an important role in infants' social and cognitive development. For example, the exaggerated musicality of IDS may stimulate infants' attention and responsiveness as well as accelerate subsequent vowel discrimination and word recognition.[21]

The wheels on the bus . . .

What makes "The Wheels on the Bus" such a fabulous song? Many parents utilize the song to calm their baby.[22] It can also be taken up-tempo to promote excitement. The repeating framework of the song's melody, rhythm, and lyrics makes it easy to learn. The vocabulary enables young children to learn the parts of the bus, so it's educational. At an even earlier age, the noises made by the moving parts—like the "swish-swish-swish" of the windshield wipers—introduce sounds that babbling infants and toddlers need to practice before they can begin to speak. The song is not only educational to kids; its never-ending structure also stretches caregivers' minds to create ever more verses. It's a terrific example of the interactive loop between infant and caregiver that results from IDS.

Facility with vowel discrimination and word recognition develops gradually, initially showing up in babies as babbling. Babbling is the hallmark of an infant acquiring the capacity to imitate novel sounds and is a key skill for developing the future ability to sing and to speak. The technical term for babbling is **vocal learning**, which refers to the ability to imitate novel sounds. Songbirds and cetaceans, aquatic mammals such as whales, dolphins, and porpoises, have this capacity. Curiously, nonhuman primates, such as chimpanzees and gorillas, do not.[23]

Infants from all cultures readily recognize lullabies as such. This is also true of adults, who recognize them better than any other type of song. A study published in 2018 investigated universal form-function associations between four different song types: lullabies, love songs, dance songs, and songs for healing. The study addressed whether there are universal attributes that make a particular song type recognizable regardless of where a person is from or which language(s) they speak. The results collected showed that 97 percent of adults from eighty-six cultures correctly identified lullabies as such. The features that differentiated lullaby type songs from the other types of songs—fewer singers (and more likely to be female), fewer instruments, lower melodic and rhythmic complexities, and slower tempos—were judged to be universal distinguishing features of the genre.[24]

A final point to consider is whether music decreases the burden of caring for young children. This is not a simple topic to address scientifically. To begin with, it's challenging to define the term *caregiver burden*—a phenomenon of the "real world" of raising children—in a way that can be studied and measured objectively outside a laboratory. When it is, it's usually in the context of whether a particular music therapy intervention can ease caregiver burden (see chapter 5). For the purposes of the caregiving hypothesis, it suffices to say, "[M]others smile and move considerably more when they sing than when they talk to infants. In fact, singing mothers commonly move in time with the music and smile almost continuously."[25] These singing mothers, simply stated, appear to be happier and more emotionally bonded to their children.

Sharing happiness

Caregivers can readily share happiness with their child by singing songs that actively engage the youngster. Consider the well-known children's song "If You're Happy and You Know It." Among other things, you can share happiness by . . .

... clapping your hands

... stomping your feet

... shouting hurray

and on and on, as far as one's imagination takes it. Children learn about the parts of their body and the actions they can do in a super-fun manner through the song. Adults have their ingenuity stretched to create new verses. It's a terrific mood lifter for kids and caregivers alike.

The origins, both place and time, of this song remain debated.[26] Some label it "traditional"; some say it developed from a Latvian folk song, others from a Latin American folk song. There are those who recall hearing it in the 1950s or early 1960s while others say not until the 1970s. Some associate the song with Joseph Raposo, one of the creators of the children's television program *Sesame Street* and the composer and lyricist of numerous children's songs, including the *Sesame Street* theme song.

In sum, that adults in all human societies use music to care for children testifies to the importance music plays in raising them. Music is the first form of communication beyond direct physical touch that caregivers use with their offspring. One of the ways infants, in turn, communicate their attention and responsiveness to caregivers is through smiling, and an infant's first smile comes in response to its mother's voice. The musicality of IDS informs babies about their caregivers' meaning and intent. As babies mature, IDS stimulates babbling, a key way in which they learn the repertoire of sounds they'll need to vocalize language and song. Later, young children learn about numbers and letters and body parts through IDS.

Infants are ready for music from birth and caregivers are ready to use music starting with their first interactions with a newborn. The more caregivers use music with their offspring, the happier and more emotionally bonded with their children they can be. The bottom line is that leveraging music facilitates caring for offspring. This increases the likelihood of survival of the child. These various examples demonstrate the power of music in caregiving to confer evolutionary benefit to the human species.

THE ENTRAINMENT HYPOTHESIS

The entrainment hypothesis asserts that music, specifically rhythm, enhances survival by enabling humans to engage effectively in synchronous activity. For a mental image of this, imagine everyone in a lifeboat paddling in unison to reach the

Nadia Boulanger, Music Teacher Extraordinaire

Born into a musical Parisian family, Juliette Nadia Boulanger (1887–1979) established herself as arguably the finest music teacher of the twentieth century. While most of her students were French, she taught quite a few Americans over the years, beginning soon after World War I with Aaron Copland. Copland credited his success in no small measure to Boulanger's instruction and they remained lifelong friends. Quincy Jones, who garnered fame as a musician and as a record producer principally in the jazz and pop genres, was another one of Boulanger's American students. Astor Piazzolla, the famed Argentinian tango composer, also studied with her. In addition to giving lessons, she hosted weekly salons that encouraged her students to interact with one another as well as meet established musicians.

Boulanger demanded compositional excellence and thorough knowledge of musical tradition, yet she also encouraged her students to explore their own personal forms of musical expression. It was in this rich and rigorous musical environment that Raposo studied. This environment, in turn, contributed greatly to his ability to create so many memorable songs, including "Sing" (1971), which premiered on Sesame Street. The Carpenters, the sister-brother duo, later took the song as high as number three on the pop chart. Its simple yet powerful message: sing your song for the lifetime of happiness and meaning it brings.

shore. By working together, the individuals meld into a team and, as a team, they are more likely to achieve their goal.

The concept of entrainment initially referred to the tendency of two oscillating bodies, repetitively moving back and forth at a regular rate, to synchronize their movements. Seventeenth-century pendulum clockmakers first described this phenomenon by noticing that after placing two pendulum clocks on the same surface, or mounting them on the same wall, the movement of one pendulum gradually synchronized with the movement of the other pendulum.[27] Such repetitive movement at a regular rate, periodicity, implies conforming to a time constraint.

Music adheres to a time constraint by definition: adhering to a time constraint is a hallmark of rhythm. For humans in the context of music, entrainment refers to the ability to coordinate motor movements to an external rhythmic stimulus, that is, a rhythmic sound source outside the body.[28] This is a very special skill that humans possess. Extending the concept of entrainment from one person's ability to coordinate movements with an external rhythmic source, the introduction of music allows people to move together in concert, coordinating their movements to the same source. The art form most strikingly linked to entrainment is dance.[29] Think of a country and western line dance with dozens of dancers executing precision steps while remaining in perfect formation (and out of each other's way).

Entrainment's multiple benefits

Engaging in synchronous activity yields two powerful evolutionary benefits. First, it is pro-social. Engaging in synchronous action leads to greater trust and generosity among people.[30] As a result, music helps to build cohesion and pro-social behaviors within a group. These benefits have been documented in multiple settings, including among young children, religious singers, and military recruits.[31] For example, after children as young as four engage in cooperative play that includes singing and playing percussion instruments, they exhibit more pro-social behavior than children of the same age who engage in similar cooperative play but without music. According to independent research on the Kindermusik program, "those children who were currently enrolled in Kindermusik (regardless of age) showed better self-regulation than those who were not currently enrolled."[32]

As another example, synchronous activity results in greater identification, trust, and generosity among participants than does nonsynchronous activity, such as observed in groups of Buddhist chanters and Hindu devotional singers compared to groups of cross-country runners. Many Western religious traditions include congregational singing as a valuable aspect of worship. People singing together is a form of simultaneous group movement. When movement in unison of other body parts—such as the hand clapping connected to gospel vocals—is added to singing, songs takes on additional dimensions of entrainment along with their additive benefits. Noting the rise in choir participation in Great Britain, two researchers at Oxford University commented that "Community singing is effective for bonding large groups, making it an ideal behavior to improve our broader social networks. This is particularly valuable in today's often alienating world, where many of our social interactions are conducted remotely."[33]

A third example relates to the military practice of marching in step. Although no longer used for mustering soldiers into battle formation, the military retains marching because of the sense of unity it fosters among soldiers from diverse political, religious, and socioeconomic backgrounds. The same objective pertains to participation in marching bands. A band and orchestra website states, "Not only does marching band exercise the mind and body, but also it encourages friendships, cultivates creativity, and provides students with a unique opportunity to grow as individuals."[34]

Plainly stated, people who make music together form social and emotional bonds with one another. As Trehub and colleagues state, "Making music together is simultaneously building a community together, which is considered by many to be the most adaptive and the most evolutionarily significant aspect of musical experience worldwide."[35]

The second powerful benefit of engaging in synchronous activity is that it is productive. Acting synchronously enables a greater amount of work to be done than can be accomplished by the same number of individuals working separately. Sea chanties—songs sung aboard ships—fantastically demonstrate this benefit. These songs—led by chanteymen, experienced song leaders who set the time and rhythm for these verses—kept the crew working in unison. Sea captains would pay good wages to a skilled chanteyman because he could get the crew to work maximally well together.

The Mystic Seaport Museum, in Connecticut, focuses on the epoch of the great sailing ships. This includes the museum's historical reenactors making use of sea chanties sung in the nineteenth century while performing duties on ships. In some cases, visitors can participate, for example, by working as a member of a group turning a capstan to raise an anchor. Anywhere between eight and sixteen people can take part. One reenactor plays the role of the chanteyman, while the "crewmembers" synchronously march in time to turn the capstan, pushing one of the multiple levers protruding from it. It's not as simple as it may seem:[36] first, it requires strength to push the levers that crank the capstan; second, it takes concentration to stay focused on the task; and third, it demands staying alert to avoid getting entangled in the numerous ropes, called lines in sailor talk, on the deck. The chantey really helps maintain mental and physical focus on the task at hand, preventing the mind from drifting off task and keeping everyone moving in unison. Thanks to the chantey, a group of strangers quickly becomes a productive team.

Sea chanties have a rich history. The word *chantey* likely derives from the French *chanter*, meaning "to sing." While some chanties were sung for entertainment, sea chanties were primarily used as work songs on merchant vessels. Maritime historian Frederick Harlow wrote that chanties were used to assist manual work of a heavy nature aboard ship: "The right song at the right time, with sailors, accomplished what otherwise could not have been done."[37] "Blow the Man Down" is probably the best-known sea chantey. Here are some sample lyrics:

> *Chanteyman:* As I was a-walking down Paradise Street
> *Sailors:* And a*way*, hey, *blow* the man down
> *Chanteyman:* A saucy young damsel I chanced for to meet
> *Sailors: Give* me some time to *blow* the man down.

> She steered me through alleys and up to the bar . . .
> *[Sailors' refrain]*
> And had me quite groggy b'fore going too far . . .
> *[Sailors' refrain]*

> She told of a clipper just ready for sea . . .
> *[Sailors' refrain]*
> And said she was waiting for sailors like me . . .
> *[Sailors' refrain]*

> And so I was shanghaied aboard this old ship . . .
> *[Sailors' refrain]*
> She took all my money and gave me the slip . . .
> *[Sailors' refrain]*

> I'll give you a warning before we belay . . .
> *[Sailors' refrain]*
> Don't ever take heed of what pretty girls say . . .
> *[Sailors' refrain]*

As the sailors sang their refrain in each stanza, they in unison pulled, or made whatever movement was required, when they came to the words in italic. The lyrics above are from Frederick Pease Harlow's authoritative book on chanties.[38] It's also likely that many chanteymen improvised lyrics, perhaps even poking fun at the ship's officers from time to time to help the crew let off some steam. Like court jesters, chanteymen had leeway to poke fun while working, whereas the other roles on board ship remained highly circumscribed and disciplined out of necessity.

How does entrainment work?

Surely entrainment, the coordination of body motion in time with musical rhythm, results from a relationship between the brain systems for hearing and movement. As previously noted, the ability to fashion rhythm results from the close interplay of auditory and movement regions of the brain.[39] Thus, it's no surprise that entrainment derives from this interaction since one entrains with rhythm. Other important aspects of entrainment are that it begins to develop early in childhood and is not dependent on musical training. This is supported by the fact that babies move rhythmically to music, and young children can readily tap to the strong beats of a rhythmic pattern regardless of musical training.[40]

The vestibular system is a robust candidate for providing the link between hearing and movement. The vestibular system is often narrowly viewed as being responsible for balance and associated with dizziness when its function goes awry, but there are strong reasons to consider it in a broader, more expansive way. The vestibular system was the first sensory system to develop in evolutionary history and is also the first system to develop in the womb.[41] That infants delight in being bounced, rocked, and swooped before they can walk demonstrates that the system matures early on.[42] Children adore the sensation of vestibular system stimulation. The popularity of amusement parks attests to this fact.

The vestibular system has multiple points of proximity to the auditory system (figure 4.2). The sense organs for these systems, the labyrinth and the cochlea respectively, are located close to one another in the inner ear. Both systems pass their information to the central nervous system—up the brainstem to the brain—via the same peripheral nerve, although the two components of this nerve function somewhat independently of one another. Moreover, as the vestibular system predates the auditory system in evolution, it was likely the first system for hearing. The neuroscientist Nina Kraus noted, "Hearing and movement have a common evolutionary origin."[43]

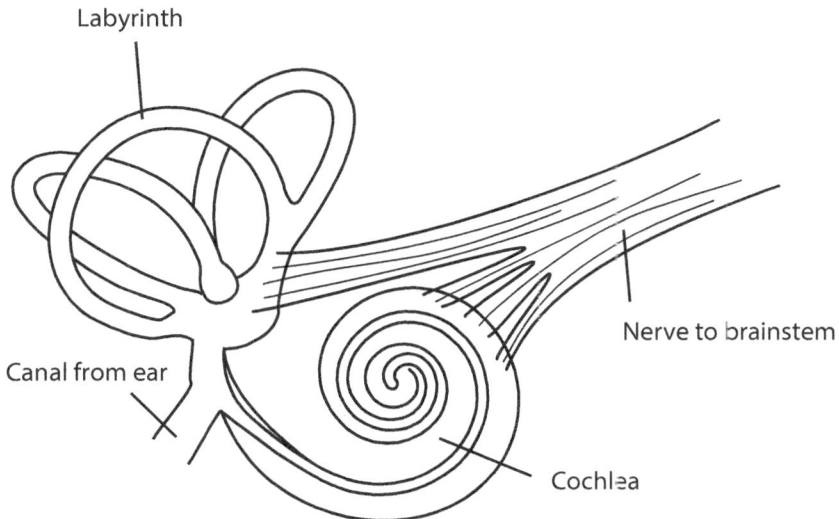

FIGURE 4.2   The cochlea and the labyrinth, showing their proximity to one another in the inner ear. Both structures serve important functions for music. The cochlea is critical for hearing sound wave information. The labyrinth plays a key role in rhythmic assessment and movement.

The idea that the vestibular system mediates rhythm dates back to 1929, to Pietro Tullio, an Italian biologist. Building on Tullio's concept, Laurel Trainor, a professor of psychology, neuroscience and behavior,[44] and her colleagues directly stimulated the vestibular system. Their research showed that the vestibular system plays a primal role in the perception of musical rhythm and that this role does not require formal musical training.[45]

Trainor and her research group studied twenty-three participants with normal hearing and diverse musical backgrounds. Each participant listened to various rhythms produced by playing drums. After that, members of the study group received direct electrical stimulation of the vestibular system while members of the control group received direct electrical stimulation at their elbows. The electrical stimulation was delivered at the same frequencies as the rhythms played on the drums. Participants in the study group could identify the various rhythms based on the electrical stimulation of their vestibular system, whereas participants in the control group could not. Furthermore, study group members reported experiencing side-to-side movement of the head even though their heads remained objectively still. That is, based on the activity of the vestibular system, they reported being aware of bodily (head) movement although they were not actually moving. Their conscious brains perceived an illusion of movement. This explains why, even if your body is still, you can feel like you're dancing (or at least moving) when you listen to a piece of music like "The Blue Danube Waltz." For some people, the sensation is so irresistible they can't stop themselves from taking the next step and physically dancing to the music. The study authors concluded that the vestibular system mediates entrainment of physical movement with rhythm.

Since entrainment involves movement, a full explanation of its mechanism requires a connection of the vestibular system with the motor system. Neil Todd and Christopher Lee, a pair of British scientists, studied this connection and distilled their findings into four key concepts.[46] The first establishes agreement with the conclusion of Trainor's group that rhythm perception occurs through vestibular activation. Second, people become consciously aware of this perception because rhythm evokes both external and internal guidance of **somatotopic** representations. What does this mean? Somatotopic refers to the sensory and motor information that enters conscious awareness being represented in the brain as maps of the body (see chapter 2). This applies both to physical body motion in space (external guidance) and to the planning of self-directed motion (internal guidance). Since rhythm activates both of these representations, a person becomes consciously aware of rhythm through actual movement or by just the thought of movement. Third, humans are born with a vestibular reward mechanism. This explains why people relish activation of the vestibular system: it innately stimulates the reward mechanism of the limbic brain. The upshot of this concept is that humans instinctively enjoy—and therefore seek (approach)—movement. This is readily apparent in children, who love to spin and make themselves feel dizzy. It becomes more subtle in adults, but the enjoyment of movement never ceases. Fourth, to be discussed in the next section, the link with the limbic system prompts the urge to dance to rhythmically compelling music like "The Blue Danube Waltz."

In short, entrainment refers to the special human ability to coordinate movements with an external rhythm. Research shows that the vestibular system mediates this ability, and it is innately rewarded—via the same limbic brain mechanism, the reward system, for enjoying music! Furthermore, the reward extends to both physical (present) motion as well as imagined (future) motion.

Entrainment enables humans to engage in synchronous movement and reap its pro-social and productive benefits. The pro-social ability to create community may well be the most adaptive and most evolutionarily significant aspect of musical experience worldwide. Music's productive aspect enables a greater amount of work to be accomplished together than can be done by the same number of individuals working separately. Both of these benefits improve the chances of survival, further evidence that music confers evolutionary advantage to the human species.

Harriet Tubman and the Powers of African American Spirituals

Some songs satisfy more than one of the hypotheses discussed in this chapter of music's power to provide evolutionary benefit. Consider, for example, African American spirituals, which the Library of Congress defines as a type of religious folksong most closely associated with enslaved African people in the American South.[47] As examples of entrainment, these songs, sung in unison or in call-and-response fashion, made slaves more productive by facilitating their ability to work together. Why, then, did slaveholders suppress the singing of them? Because the songs were also examples of communication. In fact, spirituals communicated not only hope for a better future, but also served as an enactment of marching from bondage into freedom.

One of the best-known African American spirituals is "Go Down, Moses." Some say it's about Harriet Tubman, who was sometimes called the Moses of her people. Others say Tubman used the code name Moses when she traveled to the

South to bring people out of bondage. Still others say she sang spirituals to signal her presence to help any who wanted to escape slavery.

Tubman's exploits and accomplishments are legion, and biographies about her are still being written. Yet few people know that after the Civil War until she died in 1913—nearly fifty years—she lived in Auburn, New York. This upstate town is also where I graduated high school. Tubman's home is at the southern end of town. In the 1970s, there was a marker at the house where she lived, but little in the way of activities commemorating her. In 2017, however, the home and grounds became a National Historic Park.[48]

I visited in the summer of 2022, having reserved a tour. Arriving at the appointed hour, I met people from Ohio and California scheduled for the same tour. The national historic site includes Tubman's house and its grounds, a visitors' center, and a large sculpture. The sculpture, *The Journey to Freedom,* by Wesley Wofford, stands nine feet high and movingly depicts Tubman leading a child out of slavery.[49] Tubman's house is a modest two-story structure with simple yet comfortable furnishings. The visitors' center, while small, is informative.

As I drove off after my visit, I had a sudden thought: Not only had Tubman lived there, she is also buried in Auburn. With that, off I headed to Fort Hill Cemetery. As some cemeteries are wont to be, Fort Hill is a large, beautiful expanse of land with lovely old trees providing shade to its residents and visitors. I parked my car and walked unhurriedly along the main path. I passed the graves of some people I had known. They were parents of my friends. It brought back memories of my teenage years.

I then rounded a bend and saw Tubman's grave under a full, mature tree. It seemed fitting that she should have such a lovely arbor to provide her with shade on a hot summer day. The marker identifies her by both of her married names; she was married to John Tubman before the Civil War and then to Nelson Davis after it. Her gravestone reads, "Servant of God, Well Done." This speaks for itself, far beyond any words I can add.

THE PROCREATION HYPOTHESIS

Success at finding a mate and passing one's genes to the next generation, known as sexual selection, is a key tenet of evolution. The procreation hypothesis asserts that music facilitates sexual selection. Think about it: Most song lyrics deal with love, *l'amour*, in one way or another. There are songs about falling in love and songs about falling out of love; there are "somebody's done me wrong" songs and songs for every station in between. It would be easy to believe that the reason songs even exist is to navigate the gamut of love.

Fruit of the Garden of Eden

A true story of sharing music leading to sharing a lifetime

The power of music to win my wife's heart played a major role right from our first date. This was back in the era of cassette tapes and a tape of David Broza's music was playing on my car's sound system when I arrived to pick her up. Broza is a famous Israeli singer who infuses a large measure of Spanish musical sensitivity into his songwriting. In all honesty, this cassette just happened to be in my tape deck already, I hadn't consciously thought about what music to play for the occasion. At the moment that the woman who would become my wife got in the car and we drove off, Broza's hit tune "The Woman by My Side" (1983; original Hebrew title "Ha'Isha She'Iti") began to play.

Broza sings a striking line in the song where he proclaims, "The woman by my side is the fruit of the Garden of Eden." That might be a bit much for someone to hear on a first date. Had the one by my side not also been looking for a serious relationship, she might have opened the car door and jumped out then and there. As it turned out, though, good fortune was smiling on us. Dina remarked to me months afterward that she was touched in that moment by the lyric and by the music because they revealed a window of possibility for people seeking a long-term relationship as we both were. Years later, we attended a David Broza concert and waited around after the show to meet him so we could tell him our story. He enjoyed hearing it.

It would be a mistake to think that love is only the subject of modern songs. This theme was well known to the ancients, although no records reveal how the lyrics were sung. Consider the following biblical verses, describing different facets of love, from Song of Songs:

> My beloved spoke thus to me, "Arise my darling; my fair one, come away! For now the winter is past, the rains are over and gone. The blossoms have appeared in the land, the time of singing has come, the song of the turtledove is heard in our land."[50]

> I opened to my beloved, but my beloved had turned and gone. My soul failed me when he spoke. I sought him, but found him not; I called him but he gave no answer.[51]

Ancient songs from China dating as far back as three thousand years have been preserved. In one charming song, a young maiden expresses her eagerness to find a husband and reflects upon being passed over.

> Plop go the plums
> Only 7 of every 10 remain on the tree
> All you men who want me, seize this lucky time.

Plop go the plums
Now only 3 of every 10 remain on the tree
All you men who want me, seize the present moment.

Plop go the plums
They've been gathered into baskets
All you men who want me, speak right up![52]

Lyrics of love songs also survive from ancient Egypt. "The themes of love songs are pretty universal," the Egyptologist Cynthia May Sheikholeslami says, "with the musical poetry focused on how attractive our beloved is, the longing to be with our adored, and how painful separation can be."[53]

The voice of the dove is speaking,
It is saying "The land has been made bright, what is your
   path?"
May you, bird, not scold me.
It was in his bed that I found my brother (i.e. my male
   beloved),
My heart is exceedingly glad.
We said (to each other):
"I will not be far (from you)."[54]

It's common to encounter the imagery of birds in poetry about love. For millennia, humans have regarded birds with wonder. Their capacity for flight evokes a sense of ultimate freedom, to be able to escape from any problem, to be carefree. People listen raptly to birds' chirps and calls. The birds' singing naturally awakens a sense in humans that they are seeking romance.

Songbirds

No one truly knows whether birds sing solely to meet their needs to sustain life—survival and procreation[55]—or if singing might also be an "art form" for them, something they simply enjoy.[56] Whichever it is, they appear to humans to respond to songs much like people do, so it's not surprising that neuroscientists have studied song birds extensively and learned a great deal about the neurobiology of their singing. Among the song birds, finches have garnered the most attention. Studies have shown that a young male learns and perfects the elaborate songs of his father, then performs them as an adult to attract a lifelong mate. A young female also learns the songs of her father, but she doesn't perform; rather, she analyzes every detail of a potential mate's singing and compares it with her father's example. She is very picky because the mate she chooses will help raise their young—'til death do they part.[57]

Interestingly, a song bird exposed to the singing of a closely related but different species can learn the repertoire of that species, but not master it. This has been confirmed by exposing baby male finches of one species to the adult males of another finch species.[58] This method revealed that a song bird can only perfect the repertoire of its own species. This finding is consistent with the concept that complex behaviors require the combination of a potential ability to perform the behavior inherited through genes along with the learning of the behavior through instruction and practice.[59]

The key to the brain's capacity to perfect the songs of its species is termed **vocal learning.** Vocal learning refers to the capacity to acquire new sounds via imitation. It's the ability to learn sounds heard in the environment that are not genetically coded for the individual to produce. Only a limited number of animal species have this capacity. Among primates, only humans have it. Chimpanzees, humans' closest relatives, do not have this ability; nor do gibbons, brachiating primates found in Southeast Asia.[60] On the other hand, several marine mammals, such as whales and dolphins, do have it.

This does not mean that humans are more closely related to song birds than to the great apes. Rather, it implies that vocal learning evolved independently in a limited number of animals. This is an example of convergent evolution,[61] comparable to the evolution of wings in bats being a separate process from their evolution in birds. Why vocal learning did not evolve in more species of birds remains unknown, but one researcher, neuroscientist Erich Jarvis, insightfully proposes the explanation that birds that are more verbal are easier targets for predators.[62]

Brain studies reveal song-sensitive structures in songbirds, particularly the so-called high vocal center (HVC), that are critical to their singing ability. Interestingly, research has shown that new brain cells (neurons) can be added to the HVC throughout life. This research in adult oscine (song) birds played a major role in motivating scientists to investigate the topic in mammals.[63] Thanks to the interest in songbirds, it is now known that new neurons are indeed born and incorporated into the brain circuits of adult mammals, including humans.[64] As a scientist in bird behavior, Cheryl Harding, wrote, "This research demonstrates that, contrary to what we thought in grade school, there is a lot we can learn from bird brains."[65]

Adding new brain cells to the HVC occurs in sync with songbirds' mating pattern, and, as Eliot Brenowitz and Tracy Larson have found, "the survival of new HVC neurons is supported by gonadally secreted testosterone."[66] As a result, song-related regions of the brains of such birds increase in size during mating season, but shrink (atrophy) during parts of the year when they are not needed. Why would these important brain regions be allowed to wither over the winter? "Because," as science writer Laura Helmuth points out, "brains are expensive. It takes a lot of energy to build, maintain, and fuel brain tissue. Basically, if there's some part of the brain you can do without, it makes evolutionary sense to just let it go."[67] This is also consistent with the observation by Darwin—who considered sexual selection to be the evolutionary basis of music—that mostly male birds sing, and they mostly sing during mating season.

The situation in humans is, not surprisingly, more complex, particularly since our mating season extends all year round. Nevertheless, mating behaviors that developed over eons of evolution continue to reside deep within our brain. This is a major reason why "singing . . . the act of producing musical sounds with the voice, is celebrated in every culture around the world."[68] As a result, not only is music found across the globe, people are literally drawn to musicians, especially singers.[69]

The 1960s transformed the phrase "wine, women, and song" into "sex, drugs, and rock 'n' roll." Yet the decade began the way the 1950s ended, with girls screaming and swooning for singers, among them Elvis, the Beatles, and the Rolling Stones. This changed as the 1960s gave way to the 1970s. Women's voices increasingly took center stage, singing about the complexities of deep personal relationships from their point of view. Two examples of this from the era's popular music are Linda Ronstadt and Fleetwood Mac. Ronstadt covered the gamut of emotions sparked by love, singing songs addressing themes ranging from independence ("Different Drum") to finding love at every turn ("It's So Easy to Fall in Love") to lost love ("Hurt So Bad") to not finding love ("When Will I Be Loved?").[70] Fleetwood Mac seemed to embody the era's shifting sands of relationships through their songs and through the very public turmoil surrounding the break-ups and make-ups among the band's members. One of their song that poignantly sums up the often-confusing emotional landscape of love is "Over My Head" (1975).[71] The bottom line is that these women—as well as their male singing and songwriting contemporaries—were pouring their deepest thoughts and feelings about love into music.

What about dance?

Dance plays a role akin to song in sexual selection. Consider Renoir's *Dance at Bougival*, a painting that touchingly captures this on canvas.[72] The couple sure look drawn to one another based upon the way they're holding each other. The woman, rather than placing her left hand on the man's right upper arm, wraps her left arm behind his neck, resting her hand on his left shoulder. To balance the intimacy of her embrace, Renoir paints her as coquettishly averting her partner's direct gaze. By contrast, the gentleman's eyes seem to regard her straight on and his right arm is fully wrapped around the woman's back, his hand holding her at her right waistline. Try re-creating the couple's position to see if you agree that this is a very intimate dance.[73] Beyond their closeness, the couple actually appear to be in motion on the canvas, a sense that is particularly reflected in the apparently flowing movement of the woman's gown. This sense of motion sweeps the dancers and viewers toward one another. Renoir's painting, owned by the Museum of Fine Arts Boston, has been described as "one of the museum's most beloved works,"[74] no doubt due, in part, to its dance-infused sensual power.

   This is a good time to revisit the world of birds and consider the mating behavior of the sage grouse. Generally a shy and retiring animal that prefers to hide in the forest undergrowth, the sage grouse lets loose during mating season with an elaborate dance to catch the attention of the opposite sex. Each spring, large groups of grouses assemble on sage-covered brush-land, called leks, to perform a mating ritual. They return to the same lek year after year; in fact, some leks may have been the site of the sage grouse's elaborate springtime courtship display for thousands of years.

Imagine that the sage grouses have a song, "At the Lek," based on the hit song "At The Hop" (1957), by Danny & the Juniors. In the grouses' lyrics, the males are strutting on the dance floor, while the females stand around, preening and acting coy while looking on keenly because they only want the best dancers at the lek. What's most striking about the sage grouses' courtship ritual is the way the males display. Inflating a pair of air sacs on the front of their chests, they puff themselves up to appear several times their actual size. They also use the air in the chest sacs to create a sound that can be heard from miles away. Meanwhile, they fan their tails in a manner reminiscent of peacocks. They do all of this while strutting about, hoping to impress the ladies with their good looks and their dance moves. A successful male may mate with more than a dozen females in a day. The unsuccessful fellows just go home.[75]

Music, movement, and reward

As mentioned in the entrainment section, the vestibular system plays a key part in the brain's ability to perceive and fashion rhythm by mediating the link between hearing and movement. For movement, the vestibular system executes a dual role: monitoring the performance of current physical motions of the body (external guidance) as well as imaging and planning future motions (internal guidance). In addition, activity in the vestibular system stimulates the brain's reward mechanism by way of an inborn connection to the limbic system. This is the same reward mechanism through which humans derive pleasure from music. As a result, we enjoy current and seek future movement of the body—and doubly so when movement is combined with music.

Now consider Todd and Lee's fourth concept relating rhythm to movement mediated through connection with the vestibular system: a pathway linking the limbic system with the internal guidance motor function prompts the "dance habit."[76] This means that the limbic brain's reward mechanism is the bridge between the vestibular and motor systems. The tie of the limbic brain with the vestibular system by way of the pathway for internal guidance allows for future movements to be rewarded.[77] So potent is this link that it results in a "habit," an urge that drives people to move their bodies, or at the least causes them to crave to move it. The limbic brain rewards movement, plus it rewards music, thus producing a "twofer" in the form of dance—movement entrained to music.

When did the compulsion to dance arise? Way back in evolutionary time. Todd and Lee point out that even primitive animals without an auditory system possess a vestibular system and engage in mating behaviors that involve movement. They conclude that the vestibular system itself is "a vehicle for sexual selection."[78] What does this mean to humans? It means that the internal drive to move was present in ancient animals and it's still present in us today, rewarded by the limbic brain. In addition, humans can entrain the compulsion to move to an external rhythmic stimulus. A musical rhythmic stimulus is especially compelling because the limbic brain also rewards music. This is why the desire to move is activated when the rhythm of a song is sensed, why people find themselves dancing, or at least tapping their toes, even if not consciously aware of it happening. Movement in response to music is primal. As

Copland said, "If music started anywhere, it started with the beating of a rhythm . . . so immediate and direct in its effect upon us that we instinctively feel its primal origin."[79]

Sprinkle in to the mix that an element of wanting to look good to others is naturally present in the movement.[80] Even if there may be little in the way of irrefutable scientific evidence about this in humans, anecdotal evidence abounds all around us.[81]

Unspoken Messages on the Dance Floor

A true story about the power of dance

Caveau de la Huchette is a special place for music and dancing in Paris. That's why, when my wife and I visited the "City of Light" in 2022, we made sure to include it in our itinerary. It wasn't our first time there and it was still the same wonderful place we remembered. A large mid-week crowd had come to dance long into the evening hours. The music that night was blues, but performed up-tempo, up-volume, and up-energy. Dina and I joined the crowd on the dance floor and had a blast.

We both noticed an older gentleman; to us, he looked well into his seventies. He caught our eye because he was a marvelous, smooth, and self-assured dancer. We watched him dance with one young woman after another, effortlessly leading them across the dance floor. They eagerly seemed to anticipate the opportunity to dance with him. Whether these young women knew him or were complete strangers, we had no idea, but he certainly made an impression. They formed partnerships on the dance floor—he made the women look good and they made him look good—and it sure looked like they were all enjoying themselves immensely.

So, what are the take-home lessons here? A few things.

People like to look good and move their bodies, and they enjoy doing it synchronously—entrained with a musical rhythm—in the company of another person or other people. In appropriate settings, people enjoy dancing with someone special as a couple. While not a frank mating ritual for humans, dance taps into sexual selection: a profound, beyond-consciousness desire to attract one another.

Of course, humans take many nonhormonal factors (such as intelligence, culture, personality, shared interests, and many other things) into consideration regarding their mating behaviors and choices. That said, the dance floor tells us that our "human-most" cognitive brain level is built upon deep brain levels of emotions and feelings that developed over the course of evolution beginning long before humans arrived on the scene. Those older levels continue to function, often without our being aware of them.

Music confers evolutionary benefit to the human species, which explains why we seek to engage in song and dance activities. Our brain experiences reward from music, and we feel that pleasure ultimately come into consciousness. Yet, we generally remain oblivious to the fact that when we engage with music, we are often responding to the evolutionary imperatives of survival and procreation. Keep that in mind the next time you hear someone croon a love song, watch a seductive pas de deux ballet, sing a lullaby to your child, or drop by Caveau de la Huchette to dance the night away.

CONCLUSION

Main points

- Music contributes to learning by communicating knowledge in ways that language falls short; learning facilitates the process of adaptation, bypassing the wait for a random genetic mutation.

- Caregivers are ready to use music with children from the moment children are born because music improves the likelihood that their offspring will survive; in turn, offspring come into the world ready to respond to the musicality of their interaction with caregivers.

- Entrainment is the inextricable link between music, specifically rhythm, and movement; it increases productivity (by people moving in sync with one another) and improves social relationships (perhaps the most evolutionarily significant aspect of musical experience).

- Song and dance form links to sexual selection and procreation via multiple brain networks, particularly connections among the vestibular, motor (movement), and reward systems; Darwin considered sexual selection to be the evolutionary basis of music.

A few thought-stimulating questions to ponder

- Can you think of a song or other piece of music that taught you something new?

- What songs do people in your community or culture sing to or with children?

- Can you name a song or some other piece of music that makes you feel compelled to move?

- Is there a song or another piece of music that reminds you of a time you fell in love?

CHAPTER 5

Improved Quality of Life Through Music

"I think music in itself is healing. It's an explosive expression of humanity. It's something we are all touched by. No matter what culture we're from, everyone loves music."

—*Billy Joel*

Music confers evolutionary benefit to the brain due in part to the brain's innate capacities for music.[1] Another evolutionary benefit stems from music serving as a means for learning and teaching. Through learning, a species can better adapt to its ever-changing environment; through teaching, learned knowledge can be passed on to others. If music truly contributes to learning and teaching, it should be possible to identify its meaningful effects—improved survival and improved quality of life.[2] This can be done by considering music's impact in a variety of settings.

First is to investigate whether people in normal health who engage with music have an increased life expectancy, total life span being a well-accepted marker of survival. Another yardstick is whether engagement with music enables people to enjoy normal health longer. From a scientific perspective, enjoying more years of normal health equates to a superior quality of life in comparison with those who do not enjoy normal health.

Second, among people afflicted by brain disease or illness, can music improve their condition? This is asking: can music reeducate their brains, allowing them to do what they used to do, although perhaps in new ways? This is a much more diverse set of people than those with normal health since there are many types of disease or illness that can affect the brain. This chapter will survey several brain-related conditions and the roles music plays in improving the quality life of individuals with these troubles. These appraisals will use objective data in some cases and personal interviews in others; admittedly, the quality-of-life measures cited by the interviewees are their own.

Third, for people who have illness elsewhere in the body, can music's impact on the brain assist in their healing? In other words, can a positive impact by music on the brain yield therapeutic results elsewhere in the body? This highlights the tight bond between the brain and the body, and bolsters the claim that brains serve bodies

MUSIC AND NORMAL HEALTH

It seems that articles appear on a regular basis, excitedly reporting studies touting the benefit of this or that intervention on the brain. Music's impact on the brain is no exception. For example, one reported a study showing that professional musicians who started playing before the age of seven have an enlarged corpus callosum, the information superhighway connecting the left and right sides of the brain.[3] A second study showed a direct and positive correlation between the size of the primary motor cortex, a major brain center for movement, and the intensity and duration of musical training.[4] A third article reported that the auditory cortex, a brain center for hearing, is larger in musicians compared with those in non-musicians.[5] Scans to support the findings of these studies, often showing that music lights up one or multiple part(s) of the brain, generally accompany reports of this sort.

Since seeing is believing, it's cool for articles to show—and for readers to see—pictures of the brain scans. Something crucial must be kept in mind, however: These scans are often documenting a practice effect—that is, the part(s) of the brain involved in a certain activity increase(s) in size through repetition. As described for example by researchers in Normandy, France, "Musical training results in greater grey matter volumes in different brain areas for musicians; [however,] changes appear gradually."[6] In short, this "growing bigger" process takes a long time to become measurable. Moreover, a practice effect may not translate into an effect that is evolutionarily meaningful. In other words, the presence of these physical changes does not necessarily imply that improved survival or quality of life results from them.

More significant, therefore, are the following examples of evolutionarily meaningful quality-of-life findings. A study revealed higher IQs among children engaged in music making than those who were not, with further granularity of the data revealing that children who practiced singing had a greater increase in IQ than those who played a keyboard.[7] At the other end of the age spectrum, playing a musical instrument is associated with a lower likelihood of dementia and cognitive impairment. This was demonstrated in a twin-based study wherein the sibling who played an instrument into and through adulthood was 64 percent less likely to develop dementia or cognitive impairment compared with the sibling who didn't play an instrument.[8]

Another way music making confers survival benefit is by improving impulse control and the ability to multitask. Neuroanatomical studies reveal that an area called the right frontal gyrus (RFG) is an important region for impulse control as well as for juggling multiple tasks. These studies also show that the circuitry connecting the RFG to the rest of the brain is more robustly developed in humans than in our closest relatives, the chimpanzee. A unique experiment used an interesting proxy for music: Stone Age tool making. Volunteers were taught how to make such tools, a difficult task that required, on average, more than 150 hours of instruction and practice to master. Brain scans showed the RFG to be particularly active during toolmaking among the study participants. According to one of the researchers, "Piano playing uses almost the identical network as toolmaking. You're coordinating your hands while keeping in mind all of these [multiple] goals."[9]

There's more. A 2018 study showed that music making has an age-decelerating effect on the brain, especially among those who make music just for enjoyment.[10] The authors hypothesize that making music reduces age-related cognitive decline by facilitating brain plasticity and **cerebrovascular support.** Brain plasticity refers to the brain's ability to adapt to novel information and circumstances by forming new neural circuits; cerebrovascular support refers to the richness of the brain's blood supply. A study from 2011 showed that individuals with at least ten years of music-making experience performed better on multiple tests of brain function relative to their non-musician peers ages 60–83.[11] The mechanism of this effect proposed by the authors is that musical engagement boosts **cognitive reserve**[12].

Cognitive reserve—what is it and why is it important? Cognitive reserve refers to extra or redundant cerebral pathways (brain circuits) by which the brain can access or process information.[13] This occurs slowly over time, and musical activity promotes the creation of these pathways. This makes brain function more resistant to age-related decline since aging involves the gradual loss of these pathways. The more pathways that exist beforehand, the longer the function can endure.

Music making, along with exercise and healthy diet, may reduce late life physical and cognitive decline, but here's the rub: To be optimally effective, they need to be started in or before middle age.[14] So the best time to begin is now (see chapter 6).[15]

A "MUSIC AS THERAPY" APPROACH

Music can be used therapeutically to improve quality of life for people with impaired health. Reeducation plays a large role in these settings because the goal is often to teach patients or clients new ways to do things they once did but can no longer do. Often geared toward muscle memory, this type of learning is first explained verbally, then the action is practiced repeatedly.

Music as a goal-directed intervention

Alongside maximalist and minimalist approaches to understanding music's interaction with the brain is a third approach: music as therapy, a goal-directed intervention. Music therapists are taught to help their patients or clients achieve goals by using musical knowledge and tools. Their training instructs them to prioritize patient and client outcomes above other parameters. Two observations follow since real-life outcomes are paramount: First, a music therapist's definition of music is quite fluid in a positive sense. Therapists will stretch the definitions and boundaries of music to achieve the best outcomes possible, using their musical knowledge and tools. Second, what happens inside the brain as a result of music is secondary. This is not to say that what occurs in the brain is unimportant to a music therapist—far from it; it's just to note that the clinical outcome is primary. So, fully understanding what's going on in the brain is secondary, music's primary purpose being to serve the objective of improved quality of life.

Music therapy isn't simply about a group of people getting together to have fun listening to songs. That's a caricature of a music therapist's work. Rather, it's about using music to reach quality-of-life goals, which are determined uniquely for each person or group. In the following vignette, a therapist uses percussion pieces to accomplish numerous goals: Working with rhythm will ultimately help the patient recover the ability to walk; placing blocks in both hands to bang together can get a weakened limb moving again; sitting erect in a chair to hit a drum builds strength in the paraspinal muscles (muscles running along both sides of the spine to support it); and singing improves breathing function. Music for a music therapist is, above all, a means to an end.

Observing Music Therapy in Action

The best way to get a feel for music therapy is to hear from a practitioner about a treatment session. A music therapist friend of mine related the following story one morning while we were enjoying a cup of coffee together. The patient presented here represents a composite of many patients with this condition whom this music therapist has treated.

A physical therapist colleague asked me to see a patient she had been working with who wasn't making much progress. She thought that music therapy could help him.

The patient, let's call him Carl, was sixty-seven and had recently suffered a severe stroke. His right arm and leg were weak, and he had difficulty communicating because of the stroke—he understood what was said to him, but he could barely utter a word [**expressive** aphasia]. On top of this, he appeared to be down emotionally, like many patients are after a bad stroke.

I went to see Carl and brought along some simple percussion pieces. His nurse told me that Carl had little experience with music, just a couple of years of piano lessons as a child.

I walked into the room and found Carl slouching in a wheelchair, looking glum. After introducing myself and confirming his limited musical experience, I placed a simple drum in front of his left hand.

"Please hit the drum with your left hand," I requested.

He did.

"Do it again," I said.

He did it again.

"OK, now hit it hard one time and then soft two times."

He did it.

"Keep doing it," I told him.

Soon Carl was tapping a waltz rhythm on the drum—a continuous sequence of one strong beat followed by two softer beats. A faint smile seemed to appear on the left side of his face. I knew the smile was a bit forced because it wasn't symmetrical.

"Let's try a different pattern now," I said. "Play the three beats with the same intensity and then pause on the fourth beat. I'll demonstrate: hit-hit-hit-pause, hit-hit-hit-pause." He did it slowly without a problem.

"Can you do it faster?" He gave it a good try but couldn't increase the tempo with his left hand.

"Hmm, too difficult. Let's move on to something else," I said. "Try this pattern: hit the drum on 1 . . & 3 . 4."

This rhythm is pleasing and has been used in many different musical genres: the pause created by deferring the second beat by a half count induces a sensation of movement. Famous songs using it range from Carmen's "Habañera" to "Strangers in the Night." Carl picked up the pattern quickly, and I could see his body start moving in time with the rhythm as he tapped it on the drum again and again. Working with rhythms will help his walking when he reaches that stage of his rehabilitation.

"Now," I said, "we need to get your right hand moving."

As is commonly the case, a brace had been placed on Carl's weak right hand. I put the drum aside and pulled out two wooden blocks. Each block had a strap that could be put around the hand, so the fingers weren't needed to hold the block in place. I strapped one block on his left hand, then removed the brace on his right hand and strapped the other one around it so the wood block rested in his palm. The fingers of his right hand curled up without the brace so I straightened them out and turned his hand to rest palm up on his thigh.

"Now, bang the blocks together playing the same rhythm you just played on the drum," I told him.

At first, he banged the block in his left hand against the block in his right hand, but as he kept at it, I saw him start to use his right shoulder muscles: he was trying to move the right hand toward the left since strength in the shoulder tends to be somewhat better preserved than strength in the hand in his type of stroke. I could see it was taking a lot of effort on his part, especially since he wanted to avoid hitting the fingers of his right hand when they would curl up.

"Great job," I said. "Let's move on."

I'd been bothered by Carl's posture since I entered the room and wanted to do something about his slouching. I wheeled him into the therapy gym, and with the help of his nurse, transferred him from the wheelchair to a wooden chair. I placed a table in front of him so his legs were under the table. I positioned the drum on the table at a distance that required him to sit a little bit forward in the chair.

"Let's go over those patterns you played on the drum in your room," I said to him. He happily played the waltz rhythm and the "Habañera" rhythm without a problem, not knowing that he was also strengthening his paraspinal muscles because he was sitting erect.

"I like the way you're sitting up in your chair, Carl, keep up the good work. Now let's try some singing," I said. "Let me hear you sing 'Happy Birthday.'"

It was an effort for him to form the words, but he eventually succeeded. Next, we moved on to "Twinkle, Twinkle, Little Star" and then to a simple counting song. All of them were helping prepare Carl for more intensive musical-language therapy. They were also activating his core and respiratory muscles, important not only for speaking, but also for building endurance in physical therapy.

"Carl, you've done a great job. You've earned some rest," I told him.

The nurse and I transferred Carl back to the wheelchair and returned him to his room. As I was leaving, I heard him say, "Thank you."

I turned back around to him and could see a smile on his face. It was a full, symmetrical smile so I knew that it was coming from his emotional self and meant that he was truly feeling happy.

MUSIC FOR MOVEMENT THERAPY

Music can be used to help people move better following an illness or accident. An underappreciated challenge to recovering from movement problems is the abnormalities of body sensation that may accompany them. Brain injuries such as stroke and trauma often affect proprioception, sensing where your limbs are located in space, as well as strength. Auditory information from music can provide an additional or alternative source of sensory input to assist rehabilitation.

Rehabilitating upper limbs

Music-supported therapy (MSuT) is a method for rehabilitating the fine motor skills of a weak hand and arm by playing musical instruments.[16] MSuT augments standard therapy by adding the dimension of playing music with the affected hand. This integrates the component of sound with the standard sensory-motor limb connection, adding additional (i.e., auditory) sensory input support to the rehabilitation process. In multiple trials, MSuT has proven to be effective in rehabilitating hand and arm weakness after a stroke, including positive results for improved fine motor skills in a four-week intervention.[17] Whether MSuT is superior to standard rehabilitation techniques is still debated because some researchers suspect that a patient's intrinsic motivation to engage in musical activities explains the superior improvement.[18] In other words, music-supported therapy may provide additional value beyond standard therapy techniques, but mainly for people who are already positively disposed to music.[19]

Movement sonification therapy (MSoT), an approach that transforms motion directly into musical notes, is another enhanced form of therapy with musical support. The sensory nerve pathways for proprioception and motor nerve pathways for strength are located close to one another in the brain, so brain injuries often impair both proprioception and strength.[20] MSoT makes use of music as an alternative source of sensory information since auditory pathways are located far from motor pathways in the brain, making them less susceptible to be impacted by an injury that damages motor nerve pathways. The concept is that auditory sensory feedback from music can assist rehabilitation of motor function by offsetting the lost proprioceptive sensory input.

With MSoT, movement of the arm generates musical notes that the patient hears through a speaker. These tones result in musical feedback in real-time, producing audio information about the arm's position. The same movement generates the same tones each time, so the patient learns to associate the movement with the musical sounds. This helps to compensate for deficits of sensing where the limb is located in space.[21] This is important because reduced proprioception can impede rehabilitation from limb weakness after a brain injury, such as stroke or trauma.

The authors of a study about MSoT comment that it "should motivate patients and provide additional sensory input informing about relative limb position."[22] The motivation aspect the authors mention suggests, as previously noted, that this technique may mainly help people who are already positively disposed to music.

Time will tell to what extent this form of augmented reality using sound will become a part of standard rehabilitation practice.[23]

Rehabilitating lower limbs

As described in the previous chapter, humans possess the unique ability to coordinate muscle movements with an external auditory stimulus. This property is called entrainment and it aligns with the element of rhythm in music. This alignment has led to the concept of leveraging entrainment for movement problems related to walking.

The movie *Awakenings* (1990), based on Oliver Sacks's book by the same name, serves to illustrate this concept. The film recounts the experience of a young Dr. Sacks (called Dr. Sayer in the movie) caring for a group of patients suffering from a condition called post-encephalitic parkinsonism. These patients appear to be "frozen" since surviving the encephalitis lethargica epidemic during 1918–28.[24]

A key moment in the film occurs when the young doctor throws a ball toward one of the wheelchair-bound patients. Astonishingly, the patient, presumed to be paralyzed, is able to raise his arms and catch the ball involuntarily just like a healthy person would. This leads Sacks to a keen insight: important brain circuits that initiate self-directed (volitional) actions are different from those that initiate actions performed by reflex.[25] This being the case, how can an intact yet subconscious reflex system be accessed to help people who suffer from disorders impairing their ability to initiate voluntary action? Sacks himself provided the key: "Almost all of them tended to respond in some way to music."[26] Indeed, a study showed that just the thought of tapping a rhythm can activate a key motor area of the brain.[27]

Parkinson's disease is not synonymous with post-encephalitic parkinsonism, but the two behave similarly in certain ways. This includes enhancement of movement in response to music. Difficulty walking is a frequent, sometimes incapacitating, symptom of Parkinson's disease and rhythmic musical therapy can be used to improve walking. A technique called rhythmic auditory stimulation (RAS) is helpful for patients whose gait is disturbed by Parkinson's disease.[28]

RAS is the application of rhythmic musical stimulation for the purpose of improving gait or gait-related aspects of movement.[29] In RAS, patients are exposed to auditory stimuli with a stable rhythmic pattern synchronized with the movements of walking at a selected tempo. It's a splendid example of entrainment in action. This rehabilitation method of gait training with musical support improves walking better than gait training without musical support for conditions such as Parkinson's disease.[30]

The science behind rhythmic auditory stimulation

Key thought leaders in neurologic music therapy have observed that "the [human] motor system is very sensitive to the input of tim[ing] information from the auditory system . . . Rhythmic entrainment occurs when the frequency and pattern sequence of body movements become locked to the frequency and pattern of the auditory rhythmic stimulus."[31] Moreover, entrainment is not simply a movement event occurring simultaneously with a musical event on the beat. Entrainment is a far richer interlocking than that: The brain "scales" the entirety of the movement throughout the duration of the beat interval. What does this mean?

To appreciate the significance of this concept, tap your hand against a surface once per second. If entrainment simply involved synchronizing a movement (your tap) with the beat (1 hertz), the brain wouldn't concern itself with what occurs in the time period between each tap, and entrainment would simply be a temporal (time-related) event on the beat. But the brain does concern itself with the beat interval, what occurs between each tap. As a result, it calibrates the amount of energy necessary and the optimal trajectory required to match the entire movement sequence smoothly and consistently in sync with the auditory stimulus. Per the authors of a premier textbook on music therapy, "This is extremely important in order to understand why auditory rhythm improves the temporal, spatial, and force aspects of the total movement pattern."[32] This means that RAS improves not only the timing of the steps but also the body's posture, the stride length, and the efficiency of the activation of the muscles of the legs, including core muscles of the pelvic girdle.

How much musical content prompts RAS? Put another way, since RAS is a rhythm-based intervention, is melody necessary at all or does a simple metronome suffice? I had the good fortune of working for many years with a physical therapist who incorporated music into her practice. Cynthia regularly offered music-based sessions for Parkinson's disease patients to improve their gait. Although no formal records were kept, she observed a variety of responses to music. Some patients needed nothing more than a portable metronome to upgrade their walking while others needed to hear the entire song to derive benefit. There was also a third group of patients whose response to music was limited: they required ongoing encouragement and direction from the therapist along with the melody and rhythm of a song to realize improvement in their gait. In short, there is no one-size-fits-all solution to the question of how much music is required for RAS to be helpful. The intervention needs to be individualized since different people respond differently to music.

MUSIC FOR PAIN MANAGEMENT

Pain following surgery is to be expected. Surgery, by its nature, causes injury to body tissue even in pursuit of ultimately repairing it. Furthermore, the body's own healing process involves feeling pain. Body parts affected by any type of injury, including surgery, emit pain signals received by the brain. How those signals reach and are interpreted by the brain are unique for each individual.

One leading theory of pain proposes that the brain can receive and process only a limited bandwidth of information supplied by the senses. Since pain represents just one type of sensory information, the intensity of the pain can be reduced if non-pain types of sensory information occupy part of this limited bandwidth.[33] Music is an example of a non-pain type of sensation. So, another therapeutic way to utilize music is as a form of diversion, to refocus thoughts and actions. Diversion to lessen pain helps move the rehabilitative learning process forward.

Leveraging Music to Overcome Pain

A music therapist related this case study to me, illustrating how music therapy assisted a patient in pain to improve.

The hospital where I work called me for help in getting a patient moving after surgery. The patient was a man in his sixties, quite beefy, with a prior hip replacement. He had recently undergone spinal surgery and was in so much pain that every attempt to get him moving during physical therapy ended in failure. The need to get him moving was particularly urgent because, if he didn't begin to show improvement after surgery, his insurance might no longer cover his stay in rehabilitation, consigning him to basic care only. Could music therapy help this poor fellow in some way?

The physical therapist introduced me to him. Conversing with the man, I learned that he once sang in a gospel choir so I started to sing some gospel melodies with him. With rapport established, I was able to guide him to a seated position, thus beginning to engage the muscles that support the back. Next, I brought out a tambourine, giving the patient a target to hit. Each time he moved his hand from the bed to hit it, it caused him to work his back muscles a little more, strengthening them.

After getting the man comfortable with singing, sitting, and hitting the tambourine, it was time for him to try to stand. To succeed, I started by engaging him in singing. As the two of us sang together, I gently guided him to stand. I stopped singing.

"Look, you're standing up," I pointed out to him. The man stopped singing and could barely believe it.

I then made use of entrainment through rhythm to get him walking: I sang a march song while tapping out a march tempo on the tambourine. Within a reasonable amount of time, the patient was able to walk twenty feet using a walker. With this gain, he was able to proceed with his rehabilitation program.

Music Helps a Young Man Recover from Physical Trauma

Allan's true story reveals numerous ways that music can be used in a therapeutic manner over the course of recovering from a traumatic accident.

I've known Allan for years because the woman who used to live next door to him, Edie, was my brother-in-law's mother. She

frequently showered praise on Allan, especially in the winter-
time because he cleared the snow from her driveway. Allan had
already remodeled the upstairs of his own house into apartments when he purchased Edie's house after she passed away
and renovated it into an upstairs and a downstairs apartment.
Then the house behind Edie's became available and Allan
converted it into apartments as well. All of the work Allan put
into these projects was immense. What he accomplished is all
the more remarkable because Allan is blind.

Allan suddenly lost his sight at age 24, following a motor-
cycle accident that left him hospitalized in a prolonged coma.
"I ruptured my aortic artery. Yup, I'm alive to talk about it. I
had a 98 percent chance to die, and a 2 percent chance to live
with a ruptured aorta. I truly beat the odds."

Allan credits his mother with the idea of using music to
help him recover.

"I don't know where she got this idea, but when I was in
the intensive care unit, she had the nurses play music to me
over and over again. I remember hearing Billy Joel and Paul
Simon songs when I started to wake from my coma. My nurse
told me to squeeze her hands [to signal] yes or no. Waking up
would have been even scarier without it; the music made it a
comfortable place. I think my mom wanted me to wake up in
a comfortable place."

After regaining consciousness, Allan progressed to formal
rehabilitation.

"I learned a lot about my body during my recovery from
the physical and the occupational therapists. Moreover,
"music helped me to look forward to physical and occupational therapy. It helped me to feel motivated. I heard it in
the background as I was getting ready. It got me stimulated,
especially if the DJ was very upbeat."

Meanwhile, his road to emotional recovery was no less challenging than his road to physical recovery. "When I first lost my sight, I was very depressed. I also had to figure out a way to make a living . . . [Fortunately,] music helped get me out of the bad depression I was in when I was adapting to my blindness. My mom signed me up to receive a new CD each month, and I really looked forward to getting it. I needed sensory stimulation when I was first blind, and music provided it."

Allan then related the story of how he struggled with finding an occupation after he lost his sight. "I was not an academic scholar, which was OK, because I knew I was really good with my hands and my mind mechanically. After high school, I became a licensed plumber. I excelled as a plumber. I worked hard and loved every bit of it. After losing my vision, I struggled a lot with an occupation. I tried wood working. I built Adirondack chairs, picnic tables, bird houses, and wooden canoes for the Jane Goodall Institute. I was a water-ski instructor. But none of those jobs really put food on the table so to speak."

He continued by telling the story of how he came to massage therapy. "I was in Ireland at a training camp for water ski instructors. I had a really bad sinus infection. I couldn't hear, I couldn't breathe, no vision, I was a mess. I finally went to the medical tent at the water ski site. I said, 'Doc, I need some help.' Doctor Keith said I had to wait; he was working on someone.

"Just then, a disabled sit-skier came rolling in. He said, 'Doc, I can't move my neck, it's killing me.' Doctor Keith said to me, 'Allan, go behind him and sink your elbow into his neck and shoulders.' I said, 'What?' Doc said, 'Yes, just do it.' At this point, the poor fellow is jumping up and down in his wheelchair begging me to do this—and to grab a screwdriver on the way over to jab it into him if my elbow didn't do the trick. So, I did it. I sank my elbow into his neck, then massaged his massive neck and shoulders as he was breathing a sigh of relief. Next guy comes rolling in, demanding his turn. Next thing I knew, I had a line out the tent door.

"On the flight home from Ireland, the Doc came and found me on the plane. He sat next to me and said, 'Allan, I was watching you. You're good with your hands.' He said I really should consider going to massage school. That's how it started."

Despite Allan's desire to learn massage therapy, he had to find a school that could work with his impaired sight. "I was the first blind massage therapist that my school graduated. They couldn't have done more for me. They rolled out the red carpet for me, which they did not need to do. We all learned together. The teachers learned how to work with me. I learned to work with them. It was just a great experience all around."

Now a well-established and well-respected massage therapist, Allan has a unique perspective on vision. "It's very interesting that people who have sight easily become visually distracted. But when you no longer have sight, you're able to maintain a higher level of concentration. It's almost like meditating. People who meditate can totally focus when they close their eyes. But most people have to consciously make an effort to stay within the picture, the 'video' you see in your mind's eye."

He also acknowledges the advantage of having once been able to see.

"I worked with children who were blind from birth and their minds are blank. They do not know what a palm tree looks like. They have to actually feel it. Same with how a '68

Mustang looks unless they feel it. 1 am very lucky because 1 had sight for twenty-four years so 1 have a lot of aspects of a vision library—pictures and videotapes—here [in my mind]." At times, though, he admits, "1 miss my vision. 1 kind of see again by positive energy."

Allan knew plumbing before going blind, but how did he acquire the skills to remodel and maintain apartments? "1 went to a rehab facility in Massachusetts where they work with blind people. The therapists teach people how to use a computer, how to cook, how to clean, that everyday sort of stuff. And there was a woodworking class 1 took, [taught by] this guy who was amazing. An older gentleman, he taught fencing for the blind. We learned how to use table saws, drill presses, and all these powerful tools where we could lose a finger or an arm. And we did this while listening to classical music. It was like there was a symphony orchestra on the radio in the background. We ran plywood through table saws with Mozart and Vivaldi blaring in the background."

Returning to the subject of his professional life, Allan remarks, "1 really feel that my vision loss helps me to be a better massage therapist. It's palpation. You really have to feel what's under the skin. You can't see what's under the skin, so you have to feel."

He added, "Music in the massage office is very good. It's very soothing to my clients. That helps me stay calm when trying to get a particular muscle group to loosen up. Music grounds me in a peaceful way as 1 primarily use smooth music channels."

MUSIC FOR LANGUAGE THERAPY

Returning to Louis Victor Leborgne, recall that he was the man who could only utter the sound "tan."[34] His aphasia (language problem) was studied while he was alive and then reported along with the results of his brain autopsy by Pierre Paul Broca in 1861. In scientific terms, Leborgne is the **index case** of Broca's aphasia. Broca's report of Leborgne's aphasia and brain autopsy findings greatly advanced the acceptance of the concept of localization. This is because Broca demonstrated that the lesion in Leborgne's left frontal lobe was responsible for his aphasia. To this day, this region is called Broca's area.[35]

Broca thought Leborgne's damaged language function was the extent of the problem caused by a lesion in this area. Other advocates of localization took the idea a step further, dissenting from the notion that a lesion only causes decreased function in the part of the brain where it's located. Chief among them was John Hughlings Jackson, who argued that a lesion in one region of the brain also "releases function" in seemingly unaffected parts of the brain. Oliver Sacks wrote, "He [Hughlings Jackson] saw the brain as having many functional layers, developed in the course of evolution, . . . with activity at higher levels making use of, but also restraining, the activities of lower levels."[36] As I more simplistically describe the concept, newer brain layers are built on top of older brain layers, and the older layers continue to function.

Hughlings Jackson's insight also implied something more: a region of a given brain layer can inhibit function elsewhere in the same brain layer. In particular, activity of the left (dominant) hemisphere can inhibit activity in a corresponding area of the right (nondominant) hemisphere. Put another way, when a lesion occurs in the dominant hemisphere, a previously muted function of the nondominant hemisphere can become apparent or enhanced. This is called a paradoxical release. In 1871 Hughlings Jackson published a report describing increased singing capacity among speechless (aphasic) children.[37] This led him to reason that, since music function may be released by language impairment, music function may be restrained by unimpaired language function. This kindled the concept of leveraging music to rehabilitate people with aphasia.

Melodic intonation therapy (MIT) is one of the techniques that may benefit individuals with Broca's aphasia. People with Broca's aphasia have difficulty finding and articulating words but can understand what is said to them. MIT rests on the observation that these individuals can often sing phrases that they cannot speak. MIT involves two components: melodic production (singing) of words plus rhythmic tapping by the nondominant hand. MIT thus leverages melody and rhythm, the core elements of music. The melody engages the nondominant hemisphere's musical capabilities while the rhythmic hand tapping—performed in time with the production of the words— facilitates the coupling of the sounds to the movements of the mouth, tongue, and throat.[38]

MIT is started soon after a patient arrives in the hospital. So, what's the rush since, as described in chapter 3, stroke deficits may improve on their own over the first three months after a stroke? Why not wait to see if the aphasia improves on its own before starting therapy? Well, here's another interesting aspect of paradoxical release: activity in the released hemisphere increases during those initial three months, making this the optimal time frame for leveraging MIT against the impaired activity of the hemisphere with the stroke.[39] Therefore, this technique's potential to compensate for aphasia is most promising if started soon after a stroke.

Patients who successfully complete a course of MIT exhibit improved language scores and better daily life communication scores compared with those who take standard therapy without MIT. However, MIT is a demanding program, requiring up to seventy-five sessions of ninety minutes' duration each. Consequently, not all patients with Broca's aphasia are a good fit for MIT.

When Music Therapy Might Not Be the Ticket

I distinctly recall a gentleman who, sadly, ended up not being a good fit for melodic intonation therapy (MIT).

One day, the emergency department doctor contacted me about a patient who had arrived at the hospital with a stroke causing right **hemiparesis** (weakness of the right side of his body) and Broca's aphasia.[40] The patient was accompanied by his wife as well as by his daughter.

I arrived in the emergency room to find a man in his late seventies with marked weakness of his right arm and leg along with profound expressive language difficulty. I learned that he was an accomplished scientist. His wife was lovely and supportive. Their daughter's presence bolstered both of them. Over the ensuing months, he completed standard rehabilitation and achieved the ability to walk by himself, assisted by a combination of bracing of his right leg and use of a support device that he maneuvered with his left hand. In contrast, his expressive aphasia improved very little with standard therapy.

One day he came to my office with his wife for a follow-up visit. He was terribly frustrated by his reduced capacity to communicate. Having read about MIT, I asked him to sing some simple children's songs with me. He had difficulty doing it, but I figured he might be unfamiliar with these songs since he grew up overseas. We discussed the option and, although he had never engaged actively with music, decided it would be worth giving MIT a try. I located someone familiar with the technique, and his wife booked an appointment.

I was so excited by this opportunity that I contacted the therapist after receiving permission from the patient.

"Can I come to the appointment to observe the technique?" I asked.

"Sure," she answered.

"Can I record the session?"

"Yes, if he and his wife agree to it," came the reply.

Having secured the necessary permission, I arrived to the first session with a video recorder and remained as unobtrusive as possible. I was so hopeful, and so was his wife, that therapy enhanced by music was going to help the man's speaking. Alas, it was not to be: Not having had exposure to music beyond passive listening, he simply was unable to connect to a music-based intervention.

A brilliant man, his language skills hardly progressed at all despite multiple appointments with the therapist. All involved were terribly disappointed. He had profound difficulty communicating, accompanied by tremendous frustration, until the day he passed away.

The reality is that no therapeutic technique can succeed for everyone. When someone doesn't connect with a certain method, health care providers and researchers want to understand why not, in order to learn from the experience. His wife and I both wondered whether his lack of exposure to and/or his lack of interest in music may have been factors undermining his potential success with the MIT technique.[41]

The possibility of a therapeutic technique failing to improve quality of life speaks to the importance of cognitive reserve. The more knowledge and experiences a person's brain has, the more connections it makes. The brain can then draw on that reserve when a certain ability goes awry. Without prior musical experience, it may have simply been unrealistic to expect success from a music-based intervention for this elderly man in the preceding vignette. This speaks to the importance of engaging in leisure activities, such as music, in or before middle age in order for them to be truly effective.

A number of studies have assessed what contributes to or detracts from success with MIT. Factors such as stroke size and location, varieties of melodic and rhythmic intervention, and timing of the program's onset after the patient's stroke have been discussed.[42] Curiously, a patient's prior experience with music has not been consistently considered.[43] One review that did gauge this factor noted, "Previous musical experience and training may impact the effectiveness of MIT on speech outcomes. That is, it's possible that individuals who have had musical training may benefit more from the incorporation of these musical components than individuals without musical training."[44]

TROUBLES WITH MOOD

The power of music to impact one's mood is universally recognized. Indeed, it's one of the major reasons people listen to music.[45] As one study put it, "One of the main reasons people give for listening to music is to experience or modulate their emotional state."[46] So, most people turn on some music when they need a mood shift or boost. For others, however, a change in their music signals they need to take action.

Mood Music

For my friend Vikki, music boosts her quality of life automatically, but when the music stops, she needs to take prompt action.

Vikki and I met through our shared interest in writing at an event sponsored by a publisher. At the time, she was well into writing her book as I was about to begin mine. She struck me as a together, no-nonsense woman with a great deal of wisdom and common sense. When she learned that my book dealt with music and the brain, she let me know that she had been a music major in college and that music keeps her out of depression.

Vikki was raised by a grandparent in Washington. Growing up, she describes herself as "the black sheep of the family," inquisitive and intelligent.

"I loved learning, and I loved rock 'n' roll," she told me. "I needed music all the time without knowing why."

Vikki taught herself to play the guitar and described herself as an introvert with a tendency to suffer depression. After graduating from high school in the mid-1970s, she attended an all-women's college in Missouri for a change of scenery. She made some good friends there, including a roommate with whom she remains close. Vikki majored in music at college. Janis Ian was her favorite musician during those years, and Vikki even got to meet her idol when she attended one of Ian's concerts.

Vikki acknowledges that she didn't do anything formal with her music education after college, but "I have a keyboard and a guitar for my own enjoyment."

"I'm energized when I play them," she said. "They're for recharging."

Vikki explained that music plays a critical role for her internal life guidance.

"There's always music playing in my head," she shared, but when a bout of depression is coming on, "the music stops." She knows from hard-earned experience that at that moment, she is vulnerable to sinking into a deep depression. She admits that she once considered fatally harming herself when she was a young adult. The music stopping in her head is her early warning system: her depression is about to return and she needs to take prompt action in order to deal with it.

"Very honestly, the music saves my life," she said.

I asked Vikki to describe the music she hears in her head. She answered that it's always music that she recognizes, never an unknown melody. The music can be from any genre.

"I don't get to choose what's playing, I just roll with it. It could be classical music, or rock 'n' roll, or even 'O Canada.'"

With that, Vikki proclaimed that she loves ice hockey and often attends professional games, so she hears the Canadian national anthem whenever a Canadian team is in town.

Sensing how much music has energized Vikki throughout her life, I asked for her take on why so many kids take music lessons but stop playing music by the time they become an adult. She offered three observations: (1) "Life gets in the way." People get so busy doing other things that the time for music gets lost. (2) "You have to have natural talent or a love for it." That is, a person needs to feel emotionally engaged with the music. (3) "It takes practice, but people have short attention spans." Even people with natural talent need to practice, and this requires a sizable measure of stick-to-itiveness that seems to be lacking these days.

A Word from a Music Therapist

Over the course of writing this book, I have had the pleasure of meeting Dwyer Conklyn, a music therapist who is highly familiar with the neurologic system. Dwyer earned his undergraduate degree in music and then searched for a way to harness his skills and passion. He decided upon music therapy and enrolled at Colorado State University, whose program at the time was under the direction of Michael Thaut, a key thought leader in the field and a founder of neurologic music therapy. Dwyer now practices with a regional health care network in the Cleveland area and also maintains a private practice. I asked him to describe a major challenge—that's simultaneously a major opportunity—facing the field of music therapy.

Music therapy boasts a profession that links back to the days of Benjamin Rush, with college curriculums starting in the United States by the early twentieth century, and a burgeoning national organization that coincided with many of today's "established" therapies. Despite these bona fides, the music therapy profession has grown at a relatively slow pace since the formation of that first national organization in 1950. Music therapy's biggest challenge, music therapy education, may also be its greatest opportunity.

Music therapy programs have been historically housed in universities' schools of music and designated as music degrees. While this may seem like a natural fit, one must understand some of the main admission criteria for a music program, including a musical audition and eventual proficiency with multiple instruments. This ensures that students entering a music therapy program are musicians first. Again, this may seem like a true fit, but music is a tool that a music therapist wields as part of therapy. By insulating the education of potential music therapists to primarily musicians, the field loses an important aspect of professional growth; the discourse and exchange of novel ideas from a wide range of backgrounds.

Any field will be hard pressed to sustain itself if continued innovation and ideas come from only one point of view. This is by no means intended to be a slight towards musicians. Yet, even Paul Nordoff and Clive Robbins understood the synthesis needed of multiple approaches wher they began their collaboration. However, the Nordoff-Robbins method of music therapy is currently taught to students in a music department following the same criteria menticned earl er.

Finding ways to integrate a more diverse student population into music therapy programs meets several objectives. First would be a more robust discourse and exchange of ideas, as previously cited. A by-product of this would be a normalization, or mainstreaming, of music therapy principles and ideas into more incoming professionals. Presumably, if you matriculate a more diverse student population, the program would have to share requirements with other programs, leading to more crossover classes with other majors. This continues the sharing of discourse and ideas across disciplines and introduces music therapy to non-music therapy majors. Over time, this will produce more professionals in adjunct professions who have at least some rudimentary knowledge of music therapy approaches.

Unfortunately, many universities have chosen to internalize even those classes that have historically been outsourced. Choosing to teach statistics, research, and psychology courses within the music therapy structure, with music therapy instructors, further insulates the programs and their students. For music therapy to extend and be accessible to a wider clientele, the profession must first make itself accessible to a more diverse student population.

Dwyer Conklyn, MM MT-BC
DBC3 Music Therapy
July 2024

MUSIC THERAPY FOR ILLNESS IN
YOUNG CHILDREN

Music therapy interventions for infants make use of babies' innate capacities for music. Studies of parent-infant pairs when the child is in the neonatal intensive care unit (NICU) have revealed the important, positive effects of music on well-being. The primary aim of one such study was to measure the impact of family-centered music therapy on the progress of the infant's recovery. Infants in the music therapy group improved more quickly and were able to leave the hospital sooner, on average after seventy-one days as opposed to eighty-six days for the control group.[47] The treatment group received standard NICU care plus family-centered music therapy twice a week from the twenty-first day of [the infant's] life until hospital discharge. The control group received standard NICU treatment without music therapy.

Another goal of the study was to determine whether the family-centered therapy intervention provided benefit to the babies' parents. The results revealed decreased stress and anxiety levels as well as improved mood measurements for mothers and fathers in the music therapy group, but not to a statistically significant level compared with parents in the control group. This highlights a major challenge when assessing parental caregiver burden: From the perspective of evolution, the demands of child-rearing dictate that concerns about caregiver burden yield to the health and well-being of the child. The health and well-being of the child must be the primary outcome measured by these studies with the reduction of caregiver burden only of secondary importance. That's because an intervention that reduces caregiver burden but doesn't improve the baby's outcome is not of real value from an evolutionary perspective. This explains why the child's improved health is the greatest reducer of caregiver burden.[48] Oh, the joys of parenting!

Providing Music Therapy to the Youngest Among Us

A pediatric music therapist describes ways her knowledge and skills can help young children.

Joanna Bereaud is a music therapist at a large pediatric hospital in Massachusetts. She has more than twenty years of experience and focuses on "very practical applications of music" in numerous situations.

"I often work on family support in the NICU by recording mother's voice and leaving the recording on a speaker at bedside to be played when mom is not able to comfort the baby," Bereaud said.

This addresses a common dilemma faced by a mother who has been discharged after giving birth, but whose baby remains behind in the hospital. This creates large gaps of time when the mother cannot comfort her baby. With the recording, at least the musicality of the mother's voice is available to the child around the clock, as needed.

Joanna related that a second setting for her skills is using music to assist in pain management during procedures. She adapts each musical intervention to the individual situation. Such procedures can range from changing bandages on children who have sustained serious physical trauma or undergone extensive surgeries to inserting central lines (intravenous access to large veins). She points out that even if the patient is asleep or under anesthesia, "I can play into their biofeedback displayed on the monitors," which display vital signs, such as pulse (heart) and respiratory (breathing) rates. Joanna has observed that her music can influence vital signs even when a child is asleep or under anesthesia.

A third application for her use of music is to encourage movement in children undergoing rehabilitation. These youngsters are old enough that their ability to learn through music can be leveraged to improve their quality of life. She walks in front of them, "strumming guitar in a tempo of their crippled walk but then speeding it up without notice to enable exercise and push the limits a little." Joanna said she prefers "looking at what are the possibilities not [the] disabilities."

Joanna bases her interventions with newborns on the concept of creative music therapy, a family-centered method of providing "nurturing enrichment of the auditory environment for infants" that is "infant-directed," meaning that it is "continually adapted to the neonate's [newborn's] needs."[49] Consistent with this goal, Joanna and her team work to create a "Songs & Tunes" notebook tailored to each child that includes contributions from the child's parents and is sensitive to the child's culture. This can be a very important component of the baby's care after hospitalization as it provides the parents with resources that enable them to continue music therapy at home, without the presence of a formally trained music therapist. Alternatively, the notebook can provide comfort to grieving parents if their child dies as it contains musical mementos by which to remember their baby.

MUSIC FOR MEMORY

Moving to the opposite end of the age spectrum, music can play important roles for people with dementia. While there are multiple causes of dementia, Alzheimer's disease (AD) is the most common.[50] According to the Alzheimer's Association, the **prevalence** of dementia due to AD in the United States in 2022 was an estimated 6.5 million people affected with an anticipated prevalence double that by 2050.[51]

The cognitive function most associated with AD is loss of memory, specifically of internally cued autobiographical memory, meaning at-will recall of the details of one's life. This is also known as loss of episodic memory, which is what people generally mean when they use the word memory. Although medications play an important part in managing Alzheimer's dementia, non-pharmaceutical interventions, such as music, can also play a meaningful role in caring for these individuals.

Preservation of music memory

There are two aspects of music memory to appreciate in regard to dementia. One relates to its robustness and the other to its connection with motion.

Looking at the first aspect, it's well known that demented individuals are capable of recalling songs from their past through external cuing. Why does music memory remain hardy while episodic memory falters? Because memory for music stems from sites in the brain that are different than the sites for episodic memory impaired by Alzheimer's dementia.

A multicenter study described distinct and crucial roles in music memory preservation for two important areas of the brain—one involved with externally cued autobiographic memory, the other with procedural (muscle) memory.[52] These brain areas resist the pathology of AD. As a result, these two forms of memory are spared by the Alzheimer's disease process and continue to function better than episodic memory, which is vulnerable to the AD process and deteriorates.[53]

External cueing activates music-evoked autobiographic memory (MEAM). This is why demented individuals can become animated participants when hearing a familiar piece of music despite their reduced episodic memory and the diminished capacity to interact through conversation that accompanies it. Some people with dementia appear to "come alive" when prompted by a song familiar to them.

On the other hand, internally cued autobiographic memory—volitionally recalling details of one's own life—declines in Alzheimer's dementia. This episodic memory decline adversely affects at-will recall for music, the type of memory that enables a person to remember, without external cues, what songs they listened to in the car the other day. But episodic memory is not the only way to remember a song. Semantic music memory encompasses knowledge of a piece of music (its melody, its rhythm, its lyrics, etc.) and the ability to respond to the song with appropriate emotion. External cuing activates this type of music memory.

Different structures of the brain are deployed for semantic, as opposed to episodic, memory for music.[54] Structures serving semantic memory are less vulnerable to AD than structures serving episodic memory. This is why semantic memory is generally well preserved among individuals with Alzheimer's dementia. Preserved semantic memory explains how demented individuals can hear a song (external cuing) and then participate with it accompanied by appropriate emotion. Keep in mind, though, that their ability to recollect their personal history with a song is likely to be inconsistent at best because of their diminished episodic (internal cuing) memory.[55]

As for the motion aspect of robust music memory, movement function is preserved in Alzheimer's dementia because the motor system resists AD pathology. The lost ability to encode and recall the details of one's life, a hallmark of Alzheimer's dementia, correlates with impaired function of the hippocampus, of which there are two, one on each sided of the brain (plural, hippocampi). Henry Molaison didn't have Alzheimer's dementia, but lost the function of his hippocampi after they were surgically removed (see chapter 2). As a result, his episodic memory was tragically compromised to the point that he could no longer form memories of the details of his own life following the operation. Yet, in spite of never remembering that he practiced them, Henry was able to acquire new motor skills. His motor function—including his procedural (muscle) memory—was not impaired. Likewise, in individuals suffering Alzheimer's dementia, motor skills remain intact even as dementia sets in.

Procedural memory, a capacity associated with the motor system, is the type of memory for doing well-practiced skills automatically without having to think consciously about them. Preservation of procedural memory explains why demented musicians are able to continue to play familiar pieces on their instruments, even if they have trouble remembering what pieces they played soon after the concert. So, too, dancers with dementia can remember their steps even if they have trouble recalling what they ate for breakfast. This type of memory also explains how moving to a song can enhance the recall that comes from listening to a song; both are forms of external cuing. And, thanks to preserved procedural memory, demented individuals retain the potential to learn new motor skills, such as new dance steps, even if they never remember practicing them.[56]

Leveraging music to address behavioral symptoms

Musical interventions are also useful for addressing behavioral symptoms of dementia. This is important because dementia symptoms comprise behavioral disturbances as well as impaired episodic memory.[57] Music therapy has been shown to help lessen a number of behavioral symptoms, including anxiety, agitation, and depressed mood.[58] According to Hervé Platel and Mathilde Groussard, leading music neuroscientists, "Alzheimer's patients are able to perceive and understand the emotional connotations of musical material and to react [appropriately] to its listening."[59] Moreover, this capacity persists in the musical realm even after verbal (language) skills are far gone. This is why, for example, playing a piece of music that has a slow tempo may reduce agitation more effectively than trying to "talk down" an agitated, demented person.[60]

A Grandmother's Dementia

I've long wondered if my parents could have used music to help care for my demented grandmother.

I remember the house my father's mother lived in when I was a child. It had an old-world feel to it, and the interior was dimly lit. She was able to stay in her home because a caregiver was with her around the clock.

To make my grandmother's time at home more pleasant, my aunt and my uncle lovingly decided to remodel her kitchen. Unfortunately, my grandmother in her demented state never recognized the remodeled kitchen as being hers or even belonging to her house. She became disoriented and agitated each time she entered it, fearful that she had walked into a stranger's kitchen. As soon as she exited the kitchen and stepped into her "good as old" dining room, she quickly reoriented and calmed down.

In retrospect, perhaps the situation in the new kitchen could have been ameliorated if some simple measures had been taken, such as placing familiar objects in it to help her recognize the space as belonging to her. I wonder even more whether music could have played a helpful role. Whenever she was in the kitchen, might it have been helpful to play recordings of slow tempo songs that she liked, to calm her down and enable her to feel more comfortable in the remodeled space? Of course, we cannot go back in time, so we'll never know.

Leveraging procedural memory musically

Dance is a procedural (muscle) memory activity that may benefit demented individuals who have suitable mobility and balance. Large-scale population data on this remain scant,[61] but anecdotal accounts shed a positive light on the approach.

An Irrepressible Dancer

A father with impaired memory had a love for song and dance and never missed a step.

Denise, a fitness instructor and competitive ballroom dancer, credits her love of dance to her father, John, who sang all of his life and danced since young adulthood. A bit of a loner growing up, he found himself in his early twenties serving in the US military in postwar Europe. While visiting the Coliseum in Rome, two Italian girls approached him to take their picture. He then had his picture taken with one of the girls. He got her name and address and began to write to her.

Back stateside, he decided to take lessons in ballroom dancing to complement his love of singing. He soon realized that dancing improved his popularity. His writing skills must have been pretty good too because he and Luisa tied the knot three years after their chance encounter in Rome. Whenever they had the chance, the couple would go out dancing throughout the years of raising a family.

Denise and her brother began to notice a decline in their father's memory several years before he received a diagnosis of dementia; in retrospect, he was experiencing mild cognitive impairment, a precursor to dementia. He continued dancing, however, never losing a step. His kids observed how music calmed him and excited him in turns. As Denise summed it up, "Music drove his soul."

One of John's favorite activities was attending his town's annual swing dance, where he would dance with his granddaughter.

"He was out there having a grand old time," Denise said, recalling how happy he was dancing with his granddaughter.

John's days on the dance floor came to an end during the COVID-19 pandemic as his mobility and balance declined. When he could no longer dance, he still sang, including to relate to others at his long-term care facility. It was his dancing, though, that Denise credits for keeping him as mentally and physically sharp as possible prior to his passing away.

"He heard the music, and he just had to get up and dance."

Leveraging autobiographic memory musically

Musical interventions for demented individuals can draw on autobiographic memory along two pathways, internally cued and externally cued.[62] The former pathway refers to at-will recall of the details of one's life: episodic memory. The latter pathway refers to music heard in the environment that unearths recall of the song (its melody, rhythm, lyrics, etc.) and its accompanying emotions. External cuing may or may not trigger recall of one's personal experience with the song, however, due to fading episodic memory in people with dementia.

Looking at interventions along the internally cued pathway, several groups of researchers relate that someone with dementia can better recall verbal information presented musically than when it's spoken.[63] What they are driving at is that brief texts, such as a short set of instructions, are better remembered if they are sung than if they are spoken. This reinforces the communication value of music and is similar to a phenomenon found among children, who learn better through song than through speech.

A team of scientists in Finland devised a program of musical activities including singing for individuals in early dementia. Among the results they tested was the program's effect on episodic memory. They reported that even six months after the musical training, episodic memory remained modestly improved.[64] A subsequent review of multiple studies looking at this issue suggests that the positive impact of musical intervention on episodic memory is somewhat greater than the impact of non-musical interventions.[65]

A valuable point to bear in mind is that loss of memory about the details of one's personal life leads to an impaired sense of self among patients with Alzheimer's dementia. Music, by improving their recall of details from their own life, can thus help bolster a sense of identity for these individuals.

Music's ability to stimulate the externally cued pathway attracts much popular attention. Because the brain stores memories and sentiments effectively during a music-related event, later retrieval of these memories and sentiments can be ignited by music associated with them. This is the world of MEAM, music-evoked autobiographic memory, an effect that can aid individuals suffering from loss of internally cued autobiographic memory.

People afflicted with Alzheimer's dementia retain their semantic memory of pieces of music, because semantic memory—knowledge of something without recall of personal experience with it—is centered in a network of the brain that is relatively resistant to the Alzheimer's disease process. An external cue, such as hearing a familiar song, stimulates their recall of the song; in addition, their emotions and feelings aroused by the song are appropriate thanks to rich connection of semantic memory with the limbic system.[66] It's heartening to see minimally interactive folks with dementia as if by magic brighten up and sing along to tunes they know from long ago.

Activation of the externally cued pathway sometimes stimulates episodic memory.[67] The effect is inconsistent given the impairment of episodic memory, so it's a gift when it occurs. The personal stories that music helps demented people recollect puts them in touch with their own past and may be quite interesting for others to hear. Incidentally, music that expresses sad emotions may be the most effective type for evoking recall of personal experiences so it should not be excluded from musical therapy programs for these individuals.[68]

Of what value is singing along with tunes from decades ago? What quality-of-life effects result from this type of musical intervention? According to a literature review, MEAM can be used to enhance social and communication functioning in Alzheimer's dementia.[69] So, MEAM helps people with dementia be better members of a group, reinforcing music's pro-social benefits.[70] In addition, MEAM helps lessen behavioral symptoms associated with dementia, which is helpful for caregiving.

Finally, MEAM can help restore a sense of identity to patients with dementia. This is highly significant for them and cuts straight to one of the most troubling aspects of Alzheimer's dementia: namely, that people afflicted with this disease gradually lose their sense of self. This is a consequence of losing recall of the details of one's own life. This loss of identity was evident from the very beginning of our awareness of the disease. Auguste Dieter—the patient whom Dr. Alois Alzheimer studied and reported with the condition that became synonymous with his name—was quoted as saying, "Ich hab mich verloren" (I have lost myself). [71]

Summary

Music memory generally remains robust whereas internally cued autobiographic memory declines in Alzheimer's dementia. This is because brain pathways for semantic and emotional memory remain intact whereas pathways for episodic memory, at-will recall of details from one's personal life, deteriorate. Another factor contributing to the robustness of music memory is that music-related events in one's life are well remembered in general.

In addition, since muscle (procedural) memory remains intact in AD, the potential to access music by way of the motor system persists. Trained musicians can continue to play familiar pieces of music, and learn new pieces, despite their dementia. Non-musicians with dementia can also learn new songs. It's even possible for individuals who function at low levels to use small percussion instruments in music therapy sessions.[72] Preserved motor function also helps explain the potential value of dance for people with Alzheimer's dementia.

Beyond memory loss, dementia is commonly associated with behavioral symptoms. These include basic anxiety and depression as well as issues such as agitation and aggression, which can be more challenging to manage. Music offers benefits for treating behavioral symptoms associated with dementia in a nonpharmacologic manner and may reduce the need for medication to address the symptoms.

Music may improve recall of details from one's own life among people affected by Alzheimer's dementia. Research shows that music as a therapeutic intervention can modestly improve memory along the internally cued pathway. But it's stimulation by music along the externally-cued pathway that's most striking. The latter is what's meant by a MEAM, a music-evoked autobiographic memory. MEAM can help these individuals socialize and interact. And improved autobiographic memory, by either pathway, helps demented individuals retain their sense of personal identity.[73]

This fascinating package of musical interventions in the face of dementia reveals some very special properties of music. Specifically, music is a medium that demented people can participate in even after conversation becomes difficult for them. Therefore, interaction with a demented person may well be more successful through sharing music than through dialogue. Keep these musical attributes in mind the next time you visit someone with dementia. Bring along recordings of their favorite songs, familiarizing yourself with the songs beforehand in order to stimulate communication.[74] This will allow access not only to their memory but to their very identity. Keep in mind, however, that formation of new memories is impaired in Alzheimer's dementia, so—reality check—they may not remember your visit for very long, if at all. Be gentle and realistic in your expectations.

CONCLUSION

Main points

- For people in normal health, engagement with music can improve cognitive function and lower their future risk of dementia.

- Sick infants can benefit from the musical presence of their mother's voice around the clock thanks to modern recording and playback technology.

- The combination of melodic singing and rhythmic tapping can help some people recover from a language disorder known as expressive (Broca's) aphasia.

- Music therapy can improve walking for people with Parkinson's disease by leveraging the property of entrainment.

- Music's abilities to boost mood, focus attention, and induce relaxation form a combination that can be constructively applied in many settings.

- The preservation of music memory makes music an attractive intervention in Alzheimer's dementia. Many individuals with this illness respond brightly to familiar songs, and some show improved recall of details of their life as a result of music.

A few thought-stimulating questions to ponder

- What musical actions can you see yourself taking to improve your cognitive health?

- How could your musical presence help care for children? Improve someone's mood? Reduce someone's pain or suffering?

- Is there a musical gift you could give to advance someone's recovery from an accident or illness?

- Would you consider using music to assist your parent's or grandparent's mobility, such as when getting into or out of a car or entering or exiting a house?

- Would you be willing to visit someone you know with dementia and share a recording of, or even sing, a song the person enjoys?

PART III

Sing, Dance, Play

CHAPTER 6

Coda

Adding Music to Your Life

"To live is to be musical, starting with the blood
dancing in your veins . . . Do you feel your music?"
—*Michael Jackson*

As a man is crossing the road, he sees through a window people moving this way and that, seemingly jumping up and down for no apparent reason. He says to himself, "They must be crazy," and walks on. In his haste to complete his daily to do list, he didn't pay attention. If he had, he would have heard the music playing. And then he would have understood that the people he saw moving about were, in truth, singing and dancing. Had he continued to pay attention he would have realized that he was passing by a wedding. Walking past joyous music, he missed out on the singing, dancing, and playing of music that was right in front of him. Sadly, we too often ignore it as we go about our day-to-day life.

I've written *Music Between Your Ears*, first, to convey the impact of music on the brain from the perspective of evolutionary benefit, and, second, to encourage people to engage actively with music. So much of our lives has been commercialized, turned into a commodity to be purchased for consumption. Music is no exception. Songs are now readily available on a dizzying array of platforms at the touch of a button or the tap of a screen, but that is not the way humans' relationship with music developed. For thousands of years, music was participatory. If someone wanted music, they usually had to be a part of making it. Of course, there have always been some people who were better at this than others. Still, for millennia, music was a group activity. While the talented may have taken the lead, almost everyone contributed in some way, perhaps singing in the background, dancing on the periphery, or beating a simple percussion instrument.

People's relationship to music began to change in the latter part of the nineteenth century, when Thomas Alva Edison came along. His invention of the cylinder phonograph[1] meant music could be recorded, stored, and played back at a later time. Early recording technology was of middling quality for several decades, until the 1930s, when records—flat discs manufactured from vinyl—began to appear. Their initial use was largely confined to music professionals, such as disc jockeys, but by the late 1940s vinyl records had gained wide public distribution.

My father had quite a few of those early vinyl records. They constantly crackled while playing, and the voices and instruments seemed rather distant. In their day, however, records represented a huge leap forward from anything that preceded them. People could now listen to reasonably good quality recordings on the radio or on a home phonograph if they had one. Acknowledging this as a major advancement, the renowned composer George Gershwin remarked that everyone could now listen to "good" music, by which he meant his own songs.

This came at a cost, however, as members of the general public ceased being participants in music. Instead, they became consumers of music.[2] Fewer people engaged actively with music, losing out on the power of participation. Over time, many people also forgot the importance of music, which had once been taught as a valued branch of science and mathematics.[3]

People should not fall prey to the belief that a child or an adult passively listening to even highbrow music will benefit fully from it. Attending a concert can be a fabulous experience, but being a spectator does not equate with actively engaging. Truly benefitting from music requires a degree of participation with it. As Isabelle Peretz, a leading music neuroscientist, states, "Above all, it is not enough to listen to music—it is important to *make* music to improve intellectual abilities."[4]

The Earworm

An "earworm" is a tune—typically only a fragment of a song—that plays repeatedly in one's head.[5] The tune seems pleasant enough at first, but then its persistence becomes annoying. The mystery of an earworm's origin and persistence may be rooted in music's commercialization. Some songsmiths, particularly those who compose music for commercials, are skilled at writing jingles that become earworms. The cause of earworms continues to be debated and the renowned neurologist Oliver Sacks proposed an interesting hypothesis.[6]

Sacks pointed out that until the beginning of the late nineteenth century—that is, before the invention of electronics, such as radio and recordings—people had to play music themselves or actively seek it out. Nowadays, especially with the ubiquity of personal devices, people encounter music nearly everywhere, and music is available nearly all the time. Noting that no written reference to the phenomenon of earworms exists prior to Mark Twain's description of it in the late 1870s, Sacks hypothesized that the ready availability of music today saturates the brains' ability to integrate all the music we hear into our personal life experience. As a result, some tunes just keep playing on and on in our heads.

I propose ideas and suggestions in this chapter to help people who don't currently engage actively with music start to do so. Primarily told through stories, I've grouped these ideas and suggestions into three themes: sing, dance, and play. They are meant to be simple, cost little money, and require no prior musical background. The best time to get started is now.

SING

Singing—even humming or whistling a tune—is a physical expression of melody, one of music's core components. Just about everyone can sing. Even people who judge themselves as being unable to sing can likely sing. After all, singing is innate. As noted by a team of researchers at Harvard University, "The ability to sing is evident from infancy and does not depend on formal vocal training."[7]

Being overly self-conscious, a common tendency, is the biggest hurdle many adults face in overcoming their reluctance to sing. In contrast, look at children, who, being less self-conscious than adults, feel much freer to sing. So, in what settings are you comfortable enough to let your guard down a bit and feel sufficiently self-confident to sing?

The shower is a great place to sing. Seriously. Consider the shower's advantages: First, there's no requirement to sing loudly. You have control of the volume. You don't have to wake up everyone in your building (unless that's your goal) by belting out an aria. In fact, no one need hear you since the noise of the shower can muffle the sound of your voice.

Second, showers have terrific acoustics. Opera singers have to learn how to project their voice to fill huge spaces full of different types of surfaces. Compared with vast concert halls, showers are one size, small, and they consist of one surface type, hard.

Sound waves readily reverberate off of hard surfaces, sustaining themselves well within small confines. This powerful combination of size and surface type creates rich sounds. Part of the richness comes from the full bass in the small space since a shower's acoustical properties favor lower pitches.[8]

Third, another reason voices sound better in the shower is that showers help people relax. Relaxed muscles are more flexible muscles. This allows for a greater range of motion, even for the small voice-producing muscles, enabling access to a greater range of notes. The relaxing nature of the shower itself is another reason the muscles become more supple. Soothing water pouring over the body serves as a counter-stimulation that commands the brain's attention. It's brilliant for taking one's mind off of life's troubles and reducing the physical tension in the body that accompanies them.

There are many physiological benefits to singing. Singing releases neurotransmitters and hormones associated with reward and satisfaction, such as dopamine and oxytocin, and reduces cortisol, often called the stress hormone.[9] Singing improves breathing which, in turn, improves the amount of oxygen in the blood. Even a short duration of singing can improve vascular function, with effects on heart-rate variability comparable to light-intensity exercise.[10] Plus, the core and limb muscular activity improves overall blood circulation.[11] It's a win-win all around. What will you sing the next time you step into the shower?

A friend of mine found a path into singing after he married a professional singer and musician. He joined a choir, even though he had never been actively involved with music. This added a whole new dimension to his life. Although music doesn't come naturally to him, Scott told me that singing in the choir "rounds out my brain's work," a nice thing to say to a neurologist! Singing with the group, he said, brings him joy and heals the emotional part of his brain. He observed that he can "tune out everything" when doing his choir work. A surgeon, he added that he likes to have music playing in the operating room, because fewer mistakes are made and there's better productivity when music is playing. He concluded by saying that his active participation with music "reorients my emotional make-up."

As my friend and I talked, I recalled a story from my college days. When I was a sophomore, I lived down the hall from one of the coolest kids on campus. John was a philosophy major who also worked part-time at a local gas station to earn pocket money. He would relate stories about pumping gas while discussing Plato and Wittgenstein with the customers. One time he invited me to join him and some of his friends for dinner at Victor Café, a Philadelphia landmark that calls itself the "Music Lovers Rendezvous."

The restaurant's waiters, mostly vocal performance students, take turns launching into operatic treasures while diners enjoy classic Italian cuisine. On and on this goes all evening, an aria presented nearly every fifteen minutes during peak evening hours on the weekend. The music, the food, and the wine all work together to create a delicious atmosphere. I've returned to this magical place many times over the years. Restaurants, even great ones, come and go, but Victor Café is (hopefully) forever.

It is understandable that vocal performance students pursuing singing as a career simply view standing in front of an audience as part of their job. However, reading about the history of Victor Café, I learned that when it originally opened, in 1918, and for many years thereafter, it was the customers who stood up and sang arias. Imagine finishing your antipasto, then standing up to sing "Nessun Dorma" or Carmen's "Habañera" to the other diners. And they would then do the same. It led me to realize that back then, which wasn't so long ago—the switch from customers singing to waiters singing took place in the 1970s—people had a different relationship to music.[12] People sang because they enjoyed it and wanted to share that enjoyment with others. Even if you were an amateur. Even if you weren't perfect. The patrons were deeply immersed in this musical culture and the joy they experienced from it. Victor Café gave them a platform to share that culture and joy. And sharing it was the most important thing.

Once while walking through a residential neighborhood in Manhattan, I came across a man sitting on a stoop, alone, singing operatic arias outside his front door. It was delightful to listen to him for a few moments. He had a nice voice, not great, yet solid. But it was the joy that his voice expressed as he sang that was its most captivating aspect. I also appreciated his moxie, putting himself out there, in front of every passerby who chanced along. Maybe he was out on the stoop because his family was tired of listening to him sing inside the house; that I don't know. What I do know is that for me, sensing his love of sharing music was the essence of that experience.

Novel examples of publicly sharing music arose during the COVID pandemic. Remember how difficult the social isolation was, not being able to see or get close to people for days, sometimes weeks, at a time? Some urban dwellers creatively dealt with this by singing as a socially distanced community. Opening their apartment windows, a group of people could sing with one another, either together or one after the other, and their neighbors could listen or join in.

A fun and active way to share music is by participating in a sing-along. Once the group reaches ten or more people, no one person's voice need stand out. This greatly reduces the anxiety and self-consciousness that some people fear with singing. A practical way to organize a sing-along is to find or hire a facilitator, generally someone who plays guitar or keyboard. You'll also want to select a musical theme with which everyone is familiar so they will be comfortable with it and excited to share it. When I turned fifty, I wanted to embrace and celebrate the moment in an active and shared way, so I invited a group of friends over, hired a pianist, and collected a bunch of songs from the softer side of the 1960s and '70s. We all knew the songs because we had grown up with them. We gathered in the living room and had a great time singing them. Some people shared stories, and others recalled memories that the songs brought back. Afterwards we enjoyed a champagne toast and cake.

I successfully navigated my fifties, so when I turned sixty, I figured a sing-along reprise was in order. As before, I hired a pianist, put together a similar batch of songs, and gathered a bunch of friends in my living room. Once again, we had a ball singing together. This time I also decided to put myself on the line a bit, singing a solo with piano accompaniment. I chose a bilingual Charles Aznavour number "For Me, Formidable!" (1963) for the occasion. Aznavour's music is up-tempo jazz and the lyrics, by Jacques Plante, tell of a man who longs to express his love for a woman in the languages of Shakespeare and Molière, only to discover his love is unrequited. The song cleverly switches mood in an instant yet never loses its fast, showbiz pace. I've been nuts about this tune since I first heard it.[13] Like the fellow singing on his front stoop, I thoroughly enjoyed sharing that "Formidable!" moment with my friends. So what if I was good but not great? It's the sharing that matters most.

A sing-along is an activity that brings people together. Are weddings any different? Why not combine the two? When my wife and I were planning our wedding, we decided to do just that, creating an activity to bring all of the guests together right from the start. Our experience with weddings had mostly been along the lines of the bride's side in the movie *My Big Fat Greek Wedding*: large gatherings of relatives and friends who arrive at various times so the program seldom stays on schedule, and guests talking all at once, making it a challenge to get people to quiet down so things can get started. Just substitute "Jewish" for "Greek" in the title, and you get the idea.

Prior to the marriage ceremony at traditional Jewish weddings, male family and guests meet with the groom while the women meet with the bride. This welcoming of guests is a way to get people together, albeit men and women separately. We liked the idea of having everyone together and ready when the actual ceremony began, but we wanted to update the tradition in an all-inclusive fashion. That's when we came up with the idea of having a sing-along. We found a wonderful song leader, with whom we remain friends to this day. The three of us sat at the front of the room and all of the guests sang along with us. We crooned songs from the American Songbook, from the 1960s and '70s, and some in Hebrew and some in Yiddish. As guests arrived, the room became more crowded, the sound fuller and stronger. Afterward, we all sensed that we had shared a very special moment together.

Pete Seeger, an icon of American music, first came to the public's attention as a member of the Weavers. This legendary band preserved the American folk music tradition at a time when that style of music was not particularly popular. The Weavers became a major force in kindling an American folk music renaissance that began in the 1960s. Beyond Seeger's activism in the social and political spheres, in 1996 he launched Get America Singing . . . Again!, an effort to increase public participation in music. It included two booklets bearing the project's name and containing music and lyrics to several dozen well-known songs, most of them in the folk music tradition. As indicated by the title, the goal was to encourage people to go beyond spectating . . . and sing. Leda Schubert, a children's author and Seeger biographer, wrote, "There was nobody like Pete Seeger. Wherever he went, he got people singing."[14] Seeger recorded many songs, but his greatest legacy is getting others to sing along.

Daniela Sikora directs the Ridgefield (Connecticut) Chorale. She has been involved with music since the age of three. A child of Polish and German immigrants, she credits her love of vocal music to the cultural milieu in which she grew up in Chicago. "We were poor, but we were rich in culture and joy," she told me. Daniela recalled that her father, a tenor and dance instructor, never achieved fluency in English but he passed his love of music on to her.

Turning to why people often hesitate to sing in front of others, Daniela cited American culture as a major reason, commenting that the emphasis on kids' sports comes at the expense of their participation in the arts. By contrast, Eastern European countries, such as Estonia and Latvia, maintain strong choral singing traditions. She stressed the importance of instilling a love for music in the first decade of life, but lamented that music is no longer part of the curriculum in many schools.

After graduating with classical vocal arts training, Daniela tried to make it as an opera singer. Finding she couldn't earn a living that way, she opted for a career in business. Seventeen years later, music drew her back in when she was asked to play piano for her church. This ultimately led to her association with the Ridgefield Chorale.[15] The group was about to be dissolved, she said, but she resolved to restore the chorale and has succeeded magnificently. Daniela has directed the community chorale since 1998 and remains committed to attracting singers at all levels to her ensemble. To that end, auditioning is never required to join the group. In addition, she maintains a robust, ongoing vocal education program for her singers using a combination of techniques, such as mentoring and a small ensemble program.

Daniela notes that the biggest mental hurdle many people must overcome when joining the chorale stems from having once been told—or having once told themselves—"You can't do that." To which Daniela responds, "Yes, you can."

DANCE

Dance, in most of its variations, is a physical expression of rhythm, one of music's core components. Humans have the ability to coordinate their bodily movements with an external musical stimulus: one person can do it in time with the music; two people can do it with one another, moving in unison as partners; and more than two people can move together as a group. This special, human capability is called entrainment.[16] Some authorities consider dance the highest form of entrainment.[17] And dance is something more for it allows us to experience greater joy in life because when people immerse themselves in dance, they also activate realms of emotions and feelings in association with it.[18]

Music and dance are so intertwined that they are known by one and the same word in some African languages.[19] Such is not the case in Western cultures. While lots of people participate in dance in a variety of ways, many do not. Many feel stymied by the very word *dance*. Similar obstacles that hinder people from singing also hinder people from dancing. Self-consciousness can certainly be an issue, especially since dancing is typically done even more in view of others than singing. Some people are deflated by the thought of what others might think of their dancing. As a result, people who tend toward shyness may be hesitant to dance. This is particularly true if there are experienced dancers on the dance floor, which can be off-putting to people who aren't so sure of themselves. So where can one begin? What's the shower analogy for dancing?[20]

Consider trying aerobic exercise routines to recorded music as a practical first step. You can do this in the privacy of your home as well as at a gym or studio. The workouts are geared to different intensity levels and styles, such as cardio, zumba, and steps, to name a few. When you think about it, aerobic exercise routines are a form of dance in disguise. You make specified steps, with accompanying movements of other body parts, especially the arms, in time to music. Hey, that's dance. The more you do it, the more skillful you become. You can then add flourishes of your own according to your personal taste. Plus, you get the benefits of an aerobic exercise routine: more energy with less stress, better mental focus, improved heart function and blood circulation. As a bonus, dance reduces the risk of falling.[21] It's a win all-around.

I've been doing a series of aerobic exercise programs for many years now. It takes a bit of practice to learn each routine, but once you know the steps, the movements become smooth. I then begin to add personal touches here and there and play a little game with myself to see how I can improve on the prescribed routine. Another aspect is that as the leader describes and demonstrates the workout, other exercisers or motivators perform it. This makes it seem as though you're watching line dancers, so when you perform the routine, you feel like you're joining in the line. All the while, you're exercising plus enjoying the benefits of music via dance.

Dancing to your favorite songs is great too. If you're listening to music, you can move to it, no matter what style it is.

Jazz? Certainly. Rock 'n' roll? That's an easy one. Classical? So what if you never studied ballet. Unable to stand? Tap your feet and move your legs, or tap your hands and move your arms. The specific steps themselves aren't what matters. The important thing is to move in time with the music and in a manner that's safe for you; check with your health care provider if need be. Keep in mind that you can do all of this in the privacy of your own apartment, house, or room, if that's what it takes for you to feel comfortable.

Nashville, generally recognized as the center of the country music universe, has much to offer the music enthusiast. On a guided walking tour of downtown, we visited outstanding attractions and learned that music is everywhere in Nashville, including at the Wildhorse Saloon, famous for country line dancing on what they advertise as Nashville's largest dance floor.

Line dancing is found in many different cultures around the world and is a great way to engage in dance and socialize. It's not always done in a line; some of it is in a circle or, less frequently, in other configurations. Many line dancing routines can also be great workouts. With no need for a partner, you can focus on yourself as much as you want to or need to. Of course, this is balanced by the fact that as in any type of group musical activity, you'll need to learn your part. This can take some practice. Each culture has its unique repertoire of steps for the feet combined with gestures by other parts of the body. Mastering them requires time and willingness to connect with the culture to some degree. This is perhaps why line dancing is sometimes also referred to as folk dancing.

Often the first association that pops into my mind when I hear the term *folk dancing* is dance from the Eastern Mediterranean region. I particularly think of the *misirlou*, a Greek folk dance I've seen performed many times and have joined in on several occasions. To the southeast of Greece, Israeli folk dance has a long history of enjoying broad appeal. With choreography that often emphasizes light-on-your-feet moves, it's a popular seaside activity. Originally informed by Eastern European steps, this dance form has more recently incorporated influences from many other parts of the world, such as Asia and Africa, as communities from these continents have established themselves in Israeli society.

While vacationing in New Mexico, my wife and I met a couple who turned out to live only an hour from us and we became friends. The husband, originally from Colombia, South America, introduced us to cumbia, a popular dance there. I confess that I have two left feet when it comes to cumbia, yet I see that it can be a lot of fun. It's a very flexible dance form, which can be done as an individual or as a couple or as a line dance. You can watch routines presented as line dances and then perform them in the privacy of your own residence, imagining yourself to be part of the troupe as you dance while watching the video.

Joy for many intensifies when moving as an individual within a group. Their movements become entrained with the music and with the movements of the other dancers. Entrainment is pro-social; it leads to greater trust and generosity with one another and builds cohesion within a group. Many scholars believe that building community through musical activity is the most adaptive and the most evolutionarily significant aspect of musical experience worldwide.

Music helps people to live and get along with one another. Digging into this assertion, two scientists who study social psychology reviewed a series of studies and presented "data which support the idea that music evolved in service of group living." They concluded, "People's emotional responses to music are intricately tied to the . . . core social phenomena that bind us together into groups. In sum, this work establishes human musicality as a special form of social cognition [social intelligence] and provides . . . direct support for the hypothesis that music evolved as a tool of social living."[22]

The thoughts of many people turn to partner dancing when they hear the word *dance*. Do you like to dance with a partner? This can produce anxiety for some people, so it isn't the style of dance that's right for everyone. Partner dancing requires coordinated movement. Furthermore, by design, it involves a degree of intimacy; after all, you're often touching one another. This closeness may lead to a warmer relationship and mutual pleasure. And you may lose yourself to some degree in the movement because the two move like one as they become accomplished with the steps. If you seek a closer connection through the mystery and excitement that defines romance,[23] partner dancing may be right for you.

As with other forms of dance, too much self-consciousness can be an obstacle to participating in partner dancing. One approach to overcome this is to be in a space full of strangers physically while simultaneously losing yourself mentally. Near the end of the movie *La La Land* (2016), Mia and her husband chance upon a jazz club owned by Seb, Mia's ex. They sit down and listen to Seb play the piano. He notices Mia and launches into their old love theme on the keyboard. As a result, Mia enters into a dreamy state, imagining her life if things had worked out with Seb. The dream sequence takes place in Paris, specifically at Caveau de la Huchette. By chance, on the night my wife and I saw the movie, we were planning a trip to Paris in about two months. I gently elbowed her and said, "I wonder if that's a real place." Once home, a quick internet search confirmed its existence, and we decided to check it out.

Caveau de la Huchette occupies a narrow store front along an old Latin Quarter alley, Rue de la Huchette. When we arrived, we found the walls of the entryway and street-level bar area covered with *La La Land* posters. The real action takes place downstairs, in the *caveau*, with its vaulted medieval ceiling supported by thick walls and massive pillars constructed centuries ago. Packed in the middle of the week, it turns out the place is not a sit-down jazz club as portrayed in the movie, but a stand-up-and-move dance club that bills itself as "The Temple of Swing." High-octane to say the least, the music and atmosphere are infectious. You can't sit still for very long. Dancers of all ability levels were in evidence.

My wife and I didn't know a soul there, so we were totally alone in a sense. We also knew nothing about swing dancing at the time but we let ourselves go, losing ourselves on the dance floor as we joined in. Although packed tightly, the dancers made room for one another to move, and we enjoyed ourselves immensely. When Dina and I took a break to get something to drink, we sat down at a table with a couple about our age. We commented how full of energy and people the place was in the middle of the week. "Yes," they agreed, "Paris is full of distractions.[24] There are too many things to do." Ernest Hemingway was right. Paris is a moveable feast.

Our kids are grown, we have achieved the rank of grandparents, and we now have more discretionary time. My wife and I have entered a new phase in our lives. What keeps a relationship renewed, a couple together for the long-term? While we only talked about it in the past, we're now taking partner dancing lessons to learn to dance the correct way. We'll see where it leads. Likely back to Caveau de la Huchette among other things. Life is a dancing adventure together.

Deb Linley, our teacher, has been seriously involved with dance since she was a teenager. "I couldn't imagine my life without dance," she told me. In college, she majored in performing arts and it was during those years that she met the owner of a local dance studio. Taking lessons from him, she began to compete and to teach. As the years rolled along, she took over running the business and now owns the studio herself.[25]

"We have people from all walks of life," she remarked while reflecting on her studio's diverse clientele. She estimates that about half the people who come to her studio do it for the love of dance. Although the other half like to dance, too, some other factor brought them into the studio. Often, as was the case for us, it's for a wedding they plan to attend.

Deb and I talked about why people hesitate to dance. She believes "fear of judgment," either by others or by oneself, is the chief factor. She observes that some people feel vulnerable, as if they're exposing themselves in full view of others when they're on the dance floor. This is why she makes it a priority to try to put her students at ease, telling them, "It's OK to be a beginner, and it's OK to make mistakes. The studio is a no judgment zone." She also emphasizes the importance of communicating patience to her students, both newcomers and regulars alike.

She acknowledges that operating the studio is a full-time commitment, with the constant challenge of keeping the studio in the public's mind. Still, Deb's greatest pleasure comes from seeing students advance in skill and enjoyment, watching people become more comfortable in their bodies through dancing. "Dance," she notes, "is your emotions in motion."[25]

Share the Power of Music

If you already engage actively with music, you have a valuable role to play by encouraging others to do so as well. You may sometimes feel that it's easier to slip into "performance" mode and let others sit passively than it is to try to encourage non-musically engaged folks. But the following story—told to me by a friend—shows that, with a little imagination, it can be simple and highly satisfying to get others engaged in music with you.

Friends of ours invited me and my husband for Thanksgiving dinner. They're quite a musical couple. In between the turkey and dessert, our hostess took me into their music room. She motioned for me to sit at the piano and had me playing random notes while she harmonized on the glockenspiel. Then our husbands came in, and she had mine try a cowbell. Her three-year-old granddaughter joined us, and grandpa helped her play the xylophone. Then she got all of us singing. It was lots of fun!

PLAY

Upon hearing the word *music*, many people's thoughts turn to the playing of an instrument. It's wonderful to attend a live performance and listen to skilled musicians playing. Keep in mind though that a concert-level musician is, in truth, akin to a trained athlete: someone who dedicates many hours a day to practicing in order to achieve a level of performance as close to perfection as humanly possible. That's never going to be most of us, but it shouldn't exclude us from making music. After all, you can attend a major league baseball game to watch the best players in the world and also enjoy participating in a friendly game of softball at a family picnic or with a neighborhood team.

Me, I play the piano . . . somewhat. I do it for my own enjoyment. Growing up, we had an upright piano in our house because my mother played a little classical music every so often. I started taking lessons when I was about eight years old (Thanks, Mom!) and I've pursued the instrument in an approach-and-avoidance fashion since then.

My first teacher focused on classical music, as I imagine many of her era did. Simplified arrangements of Bach, Beethoven, and others were the order of the day in the mid-to-late 1960s. Yet tons of great music were being written then and, contrary to what my parents thought, it wasn't all heavy metal rock 'n' roll. This era also produced lots of terrific "softer-side" music, the kind I wanted to play. In retrospect, it seems as though the aim of piano teachers back then was to make concert pianists out of all of their young students, but that was an impossible target to hit. Maybe piano teachers during their entire career will come across a few students who become concert-level pianists. What's the goal for the rest of us?

During medical school, I shared a house with three other guys. One of them, Jeff, was a bona fide piano player. He told me that he had to decide between going to medical school versus becoming a concert pianist. He chose the former mainly because of lifestyle issues. Concert pianists are frequently on the road, performing one night here one night there. It can be a lonely existence. While we were in school, Jeff purchased a beyond gently used upright piano and had it delivered to our living room. He would come home exhausted after a long night on call, sit down at the piano, and play pieces like the entire first movement of Beethoven's Fifth Symphony—solo piano, completely by heart. He could make that tin-can of a piano sound like he was on stage at Carnegie Hall. Not only that, after twenty or so minutes of nonstop playing, he was totally refreshed, like he had just woken up from a nap. I was mesmerized.

One time when I went to visit my parents, I fished some sheet music out of the piano bench and took it to our group house. When Jeff wasn't at home—I would have been way too embarrassed to have his ears hear my playing—I practiced several songs I had once played. It was a joy to play them once I re-learned them. Playing them also helped get me through the difficult days of my father's illness and death.

After medical school, life got in the way again. I went my way, and Jeff and the piano went theirs. I barely touched a piano until my mother downsized and shipped her upright to me. The instrument sat in our living room, full of kids' toys, for a year. I then decided I had to do something with it. By chance, the mother of one of my sons' friends was a talented pianist. She agreed to take me on as a student, on a once-a-month basis. My only condition was that I could choose the songs I wanted to learn. No problem. Although classically trained, she could play and was familiar with a wide variety of musical styles.

Her incredible musical skills were driven home to me the day I played Gershwin's "I Got Rhythm" for her. I had practiced the piece for a month, and felt proud that I could play it. After I'd finished, my teacher's first comment was, "Early in the piece you played a D-natural, but it should have been a D-flat." I was flabbergasted that she could notice such a little thing, and yes, I should've played a D-flat. The episode made me realize how incredibly gifted she was. Actually, most of our conversations were delightful, talking about music on an adult level regarding what and how composers were trying to communicate. Ultimately, as her kids and mine became teenagers, she stopped giving lessons, so I had to find a new teacher. This is when I met

Paul Nickolloff, with whom I've been studying ever since and whom I describe in the preface.

Enough with my story, let's talk about yours. Perhaps you don't presently participate in music, but now you're thinking you might like to? Maybe you took music lessons as a kid, or maybe you didn't. Either way, maybe the thought of taking up an instrument at this stage in your life, whatever that stage may be, is exciting but also daunting? After all, the majority of instruments require years to master. If only there was an instrument that you could play reasonably well from the first time you tried it and that required little in the way of practice. Might that entice you to give it a try? Fortunately, there are a few such instruments. You are by now familiar with the primacy of rhythm and melody in music, so I've chosen an instrument representing each element to tell you about. Let's take a look at drums and bells.

Drums

For rhythm there is nothing more foundational than a drum. Some drums can be difficult to play while others are easier. A group of people can come together and form a drum circle. When guided by a skilled facilitator, drumming can be modified to the extent that even someone with no prior experience can enjoy doing it.

My friend Joel had no musical experience when he joined a drum circle several years ago. He was searching for a social activity involving music and, in addition, thought drumming could be helpful for rehabilitating his right arm, which was partially paralyzed from an injury. His drum circle met periodically, and he felt that lack of musical experience was not a barrier to participating. In fact, he went so far as to say, "Anyone can do it." His group's facilitator introduced rhythms from different musical traditions, some more complex than others. Joel wasn't able to do some of the more complex ones, due to his lack of experience coupled with the weakness of his right arm.

Joel enjoyed his time with the group and noted four particularly positive aspects of joining a drum circle: First, drumming is accessible; people can actively participate at vastly different levels of ability or experience. Second, the drum circle is a very social activity and thus builds community. Third, his drum circle was held outdoors (weather permitting) making it a very healthy pastime. Fourth, being exposed to and learning new types of rhythms provides mental as well as physical benefits. Joel mentioned that there were percussion instruments other than drums, such as gourds of various types, available at his circle. In addition, he noted that incorporating metal jingles added an extra dimension, turning a basic drum into a tambourine.

Joel also took lessons from a teacher to improve his technique. He put me in touch with Dave, who has played drums professionally for decades, with credits ranging from classical orchestras to jazz bands. Dave is especially drawn to jazz improvisation because he enjoys being part of the creative process of making music, which he described as "magical." He got involved with drum circles because his wife was having trouble using her left hand due to an illness. "Here, hit this drum," he would say to her. After realizing that hitting the drum was helping her, he approached a local hospital to sponsor a drum circle as a form of combined physical and emotional therapy for patients. The collaboration was a success, he said, adding, "Some of the people were angry about their diagnosis and could take out their anger on the drums." This confirmed what he had observed over many years—"Music has the ability to directly affect your emotions."

Later, Dave got involved with drum circles in a more general way, drawn to their positive attributes, beginning with the drum's accessibility. "Everyone can participate," he said. "All you have to do is hit it." Following on this was the gratification of being a musical participant: "You are contributing in a positive way to the success of the entire [percussion] section." Next came the physical and emotional benefits noted above. Finally, there was the centrality of rhythm. Dave referred me to the Rhythm Research and Resources website, which states, "Drumming is as fundamental a form of human expression as speaking and likely emerged long before humans even developed the capability of speech as a form of communication."[27] The site also notes, "The first sound we ever heard while still in our mother's womb was the beating of her heart and the rhythm of her breath. Rhythm is thus a common experience that exists for all human beings."

Bells

Tone bells are a simple instrument for playing a melody. The term *bells* is a bit of a misnomer; they're not really bells at all. Rather, the instrument consists of a series of metal plates, each tuned to a specific note, much like a xylophone. Also, as with a xylophone, they are played by striking the plates with a mallet. The simpler the instrument, the fewer plates it contains. For example, a simple set might have enough plates to form one octave. This comprises C-D-E-F-G-A-B-C (do-re-mi-fa-so-la-ti-do) for the C-major scale. It can be further simplified to a pentatonic (five note) scale. More advanced sets may contain two octaves of plates, and some sets may offer plates for other keys as well.

At the simplest level, one mallet can be used, since melody is a series of single notes. Two mallets can also be used, one held in each hand. There are a couple of reasons to use a mallet in each hand, and they relate to producing quick notes and slow notes. For quick notes, the reason is intuitive; by holding a mallet in each hand you can reach more notes per unit time. For the slow notes, the reason is less intuitive and goes to the issue of sustaining. Slow notes refer to notes that need to be held, or sustained, for an extended period of time. In a wind instrument, this is accomplished by maintaining a steady air flow, and with string instruments, by maintaining steady contact of a bow drawing across the string. In the case of a striking (percussion) instrument, this is not possible, because the strike cannot be maintained; the strike is a rapid action, and the tone will be muffled if the mallet remains in contact with the plate. A tried-and-true method for sustaining a note on a percussion instrument is a little trick called a trill, which entails playing the target note and the note just above it in rapid succession for the duration of the sustain. This rapid succession of notes can only be accomplished using two mallets.[28]

Hand bells, by contrast, are actually bells. They are similar to tone bells in the sense that each bell is tuned to a specific pitch. A bit of practice is needed to perfect the technique of ringing the bells, but it's straightforward and quickly mastered. Since a single person can only manage a few bells at a time, hand bell music is played by a team of people. The teamwork required among the hand bell ringers to play a piece of music can be a highly bonding and rewarding experience.[29]

Hand bells can be rung in various ways. The most basic method is to "shake" it, so the clapper strikes the inner surface of the metal. The bell can also be placed or held in a stationary position and the metal struck from the outside. For example, a bell can be laid on a table and struck with a mallet. Interestingly, bells of different octaves require mallets of different materials to sound the bell effectively; the higher the octave, the harder and denser the mallet material needs to be. One final method of sounding a bell is to skim a dowel around the base of the bell to create a "singing bell."

Chimes, which resemble large tuning forks, produce another type of sound. Ringers—people who play chimes and bells—describe the sound of chimes as "purer" because proper tuning forks don't produce overtones. Like bells, chimes can be played by a movement of the hand or struck with a mallet.

Rick Wood, director of the group Chime In!, emphasizes how hand bell music brings people together.[30] "Our group has ringers ranging in age from 8 to 80," he told me. He also seconded the notion that no prior experience is necessary to join and enjoy playing with a hand bell group, saying, "One of our ringers had started with the group just seven weeks before [playing in] a recent concert."

Rick related that the COVID-19 pandemic took a toll on his organization. A sizable number of people indoors at close quarters is usually required to make this sort of music, but social distancing rules prevented the group from getting together for months. Also, the bells and chimes are too numerous and too expensive to be taken outdoors. On the other hand, this encouraged small groups of ringers to form, which developed a different take on hand bell music.

Hand bell ensembles are particularly popular in schools and churches, Rick explained, because most music for hand bells is composed to be played in those settings. It doesn't have to be that way, however. A few hand bell ensembles play music geared more toward a general audience.

Adding a drum circle to a hand bell ensemble would create a group of people playing rhythm along with another group playing melody. Imagine the possibilities!

CONCLUSION

This chapter is an invitation to engage with the power of music and offers three paths for actively participating in music if you don't currently do so.[31] I encourage experimenting to find the path true to who you are:

- Path 1 (sing): This is a great route to take if you enjoy expressing yourself vocally. There are thousands of great songs from which to choose. Start by singing them in the shower and see where the path takes you. After a while, you may even get comfortable with the idea of sharing your music with others and support others wishing to share their music with you.

- Path 2 (dance): If moving your body is the way you like to express yourself, then this is the road for you. Many beginner options exist for this ranging from straightforward dance sequences to exercise routines set to music. Then personalize the routines once you feel comfortable with the steps and the enjoyment of feeling your body in motion. Who knows where it will lead?

- Path 3 (play): If you once dabbled in an instrument and imagine returning to it, or if you dream of at last playing a musical instrument, this is the path for you. A large number of adults took music lessons at some point while growing up. Half stopped playing by the end of high school,[32] and fewer than 10 percent play an instrument beyond the age of thirty.[33] This means that many adults who don't play an instrument today took lessons as kids. Surely, more than a few have asked themselves, "Why did I ever stop?" If that sounds like you, this path provides some simple and accessible options to help you resume playing. For others, it offers a fresh start.

A FINAL THOUGHT

How can you avoid veering off your chosen path? Truly, it's no problem as long as you follow this sage wisdom: "Make your own kind of music."[34]

Afterword
Scott C. Shuler, PhD

Before children talk, they sing.
Before they write, they draw.
As soon as they stand, they dance.
Art is essential to human expression.
—*Phylicia Rashad*

Samuel Markind's book is an enlightening and potentially powerful contribution to the fields of music and education. The rationales advanced for studying and making music have pragmatic importance for music educators, not only because they justify placing music within the core curriculum of schools and allocating a portion of students' school days to music study, but also because those rationales should guide the "how" and "what" of music teaching.

Throughout time and across civilizations humans have reveled in making music, while also marveling at and speculating about the reasons for music's power. Countless philosophers, theologians, educators, and aestheticians have advanced theories about how music works its magic. Music has been variously characterized as developing the soul, regulating emotions, cultivating (or undermining) morality, sharpening various aspects of the intellect, being a discrete intelligence in its own right,[1] strengthening the vital organs, putting humans in touch with the divine, providing metaphoric insight into the ebb and flow of life experience, and even harmonizing the universe.

Markind's book adds an important new perspective, one supported by *science*. Drawing on evidence uncovered by a wide range of objective research, he makes a compelling case that music's ineffable impact has its roots in human biology—indeed, is part and parcel of our central wiring—and that exposure to and involvement in making music is a fundamental aspect of human functioning that originates in, stimulates, elevates, heals, and even contributes to the evolution of the brain.

The ancient Greeks—and hence, later, the Romans—recognized the power and importance of music by including it in the Quadrivium,[2] which was the core of their education system. They sought to expose students to songs that elevated their minds and souls. Plato believed that music—the definition of which then included what we now differentiate into melody, poetry, and dance—was important "because rhythm and harmony penetrate deeply into the inmost soul and exercise strong influence on it."[3] Looking back with the advantage of contemporary biological and cognitive science, the Greeks' use of the word *soul* might be considered their attempt to label what we now characterize as the mind, or perhaps human consciousness, which—as Markind points out—can be engaged and enlivened by music and music making.

Because music can express and foster profound emotion, including a sense of ecstatic transcendence,[4] it has often been viewed as a gift from one or more deities. Music has served as a means of honoring and connecting to the divine, playing a key role in worship across civilizations throughout recorded time. Speaking personally, I rarely feel myself open to the spirit of worship in a service after hearing words alone; the experience of well-performed music, on the other hand, almost always does the trick.

In fact, the impact of music is *so* strong that some religious groups have severely limited the types considered permissible or even banned music (and its natural partner, dance) altogether, fearing that their inherent appeal is so compelling that they might draw worshippers' attention away from their deity. Saint Augustine addressed this dilemma in his *Confessions*:

> "[I am] inclined . . . to approve of the use of singing in the church, that so by the delights of the ear the weaker minds may be stimulated to a devotional frame. Yet when it happens to me to be more moved by the singing than by what is sung, I confess myself to have sinned criminally."[5]

Music is also a powerful cultural and political force, which authoritarian leaders over the years have attempted to stifle by restricting content and/or access. Protest songs have effected change, but other genres can do so as well:

> Authorities in the Russian republic of Chechnya have banned all music that is either too fast or too slow. Chechen Culture Minister Musa Dadayev said last week that all musical, vocal, and choreographic compositions must have a tempo between 80 and 116 beats per minute, in keeping with Chechen musical tradition. The directive, part of a government campaign to suppress Western influences in the conservative, mostly Muslim region of 1.4 million people, effectively outlaws most pop, including global megahits.[6]

Based on the evidence Markind presents, it seems that humans' biological urge toward music is powerful enough to defeat any such efforts to rein it in. As Andrew Fletcher famously said, "Let me make the songs of a nation, and I care not who makes its laws."[7]

Far from fearing music and the other arts, philosophers—particularly those in the specialized subfield of aesthetics—have been intrigued by and attempted to explain their influence and appeal. Aesthetician Suzanne Langer theorized that music is a dynamic metaphor for what she called the "life of feeling." She identified instrumental music as the art form most conducive to profound aesthetic experience, because it is the most abstract form, i.e., the form least obscured by an intervening object or language. She speculated that the power of a piece of music over us is linked to the extent that it parallels the flow of our personal life experience:

> The tonal structures we call "music" bear a close logical similarity to the forms of human feeling—forms of growth and of attenuation, flowing and stowing, conflict and resolution, speed, arrest, terrific excitement, calm, or subtle activation and dreamy . . . Such is the pattern, or logical form, of sentience; and the pattern of music is that same form worked out in pure, measured sound and silence. Music is a tonal analogue of emotive life.[8]

Markind writes about the ability of particular pieces of music to evoke memories and feelings associated with earlier occasions when those pieces were encountered. Perhaps future research will identify resonance between the imprints on the brain of past experience and music's ability to activate particular emotions linked to those events.

From the earliest days of American music education, practitioners have recognized multiple positive outcomes of music study and cited them in their advocacy efforts. Music first made a significant entry into public school curriculum in the United States in 1838. Lowell Mason, often referred to as the father of American music education, enlisted other influential Bostonians to convince the Boston Board of Education to make vocal music study part of that city's required core curriculum. Although the supplicants' rationale was—for obvious reasons—not based in brain science, they too anticipated elements of Markind's thesis by rooting their arguments in music's positive influence on the intellectual, moral, and physical development of students.[9]

Prior to beginning doctoral study at the Eastman School of Music in 1983 I had already encountered research that identified a wide range of positive impacts of music on children. Those outcomes often highlighted music's "instrumental" influence, i.e., increasing learning in *other* content areas. This frustrated many music educators, who know—from personal experience and observation—that music has *intrinsic* appeal and value for children. They therefore resented having to justify its place in the core curriculum by citing research indicating that making music had a positive impact on *extrinsic* elements such as spatial reasoning, standardized reading scores, and school attendance. Furthermore, in the absence of a clear scientific explanation of why students in ensembles outperform their peers academically, or why elementary instrumentalists who are "pulled out" of their classes for group lessons perform just as well on content tests as the students who remain in class, school administrators tended to dismiss claims that music and music-making yield extramusical benefits.

Hence, I vividly recall being confronted for the first time by compelling objective evidence of music's direct impact on our brains. At Eastman I attended a presentation by Dr. Frank Wilson, then chief of neurology at the Kaiser-Permanente Medical Center in Walnut Creek, California and assistant clinical professor of neurology at the University of California School of Medicine in San Francisco. Wilson presented color electroencephalogram (EEG) slides comparing the locations and intensity of brain activation in someone solving mathematical problems or reading text vs. someone performing music. Math and reading activity each generated glimmers of color in two or three locations of the brain, whereas during musical performance both hemispheres of the musician's brain lit up like a Christmas tree. (Markind quotes Hervé Platel as referring to this phenomenon as "a veritable 'neural symphony' in the brain.") In a 1986 *New York Times* article Wilson is quoted as saying, "'You can't look at the human brain without saying this is the brain of an organism designed to interact with its environment in a musical way."[10]

Around that same time one of my dissertation advisors, Dr. Donald Shetler, was conducting research on whether and how fetuses respond to music. He placed headphones on the stomachs of pregnant volunteers and played selected musical pieces at a volume likely to enable the fetus to hear. Consistent with the findings of other researchers, Shetler discovered that post-partum infants who had listened to particular musical selections while in utero reacted very differently to those pieces after birth than they did to music to which they had *not* been thus exposed, and also very differently than other babies who had not gone through the in-utero treatment. In other words, within a surprisingly short time after conception humans are already wired to respond to music and retain what they have heard.

> A well-known conductor once told a story that illustrates the above phenomenon. He was preparing to accompany a cello concerto that he believed to be unfamiliar, because he knew that he had never performed the work previously and did not recall ever listening to it. However, the work proved hauntingly familiar. It turned out that his mother, a professional cellist, had prepared and performed the work while pregnant with him. The conductor concluded that his familiarity with the work was the result of a pre-natal memory, because his mother's stomach had been in contact with her cello and transferred the vibrations to him in utero.[11]

The *Conceptual Framework* for the 2014 National Standards in Arts Education presented five Philosophical Foundations or rationales for why every student needs an education in music and the other arts:

1. *The Arts as Communication:* In today's multimedia society, the arts are the media, and therefore provide powerful and essential means of communication. The arts provide unique symbol systems and metaphors that convey and inform life experience (i.e., the arts are ways of knowing).

2. *The Arts as Creative Personal Realization:* Participation in each of the arts as creators, performers, and audience members enables individuals to discover and develop their own creative capacity, thereby providing a source of lifelong satisfaction.

3. *The Arts as Culture, History, and Connectors:* Throughout history the arts have provided essential means for individuals and communities to express their ideas, experiences, feelings, and deepest beliefs. Each discipline shares common goals, but approaches them through distinct media and techniques. Understanding artwork provides insights into individuals' own and others' cultures and societies, while also providing opportunities to access, express, and integrate meaning across a variety of content areas.

4. *Arts as Means to Well-Being:* Participation in the arts as creators, performers, and audience members (responders) enhances mental, physical, and emotional well-being.

5. *The Arts as Community Engagement:* The arts provide means for individuals to collaborate and connect with others in an enjoyable, inclusive environment as they create, prepare, and share artwork that brings communities together.[12]

As someone who helped develop both the National Standards and their *Conceptual Framework*, I find it heartening that Markind's presentation of the role and nature of music supports those philosophical foundations.

Musicians and educators will also be interested in Markind's discussion of the key elements of music, which draws on a blend of neurological research and Aaron Copland's writings. Music educators will find affirmation of current instructional methodologies in Markind's identification of melody (a pitch sequence) and rhythm as the fundamental building blocks in how the brain responds to music. All of the most popular approaches to teaching music at the elementary level—Gordon's Music Learning Theory, the Kodály and Orff-Schulwerk approaches, and Feierabend's Conversational Solfège—are largely organized around teaching those two dimensions. Markind also makes passing reference to the importance of harmony and timbre. Apparently, further research is needed into the role of dynamics—a.k.a., loudness and softness—in listeners' response to music. Performers know that the expressive shaping of dynamics contributes to the rise and fall of intensity, and hence listeners' expressive response.

Continuing on the topic of expressive or affective response, it is informative to compare the comments of Markind's imaginary music neuroscientist on syntax and rewards to aesthetician Leonard Meyer's Theory of Expectations. Both emphasize that affective or aesthetic responsiveness requires familiarity with syntax for the particular style of music being experienced, including the ability to make predictions based on that syntax. Drawing on the work of Robert Zatorre, Markind emphasizes the *qualitative* dimension:

> if the next sound turns out to be better than what was predicted, the reward system will respond with increased dopamine release, the currency of reward in the brain.

Meyer, on the other hand, emphasizes the *accuracy* of prediction based on the listener's syntax. According to his theory, intensity or tension is created by inaccurate predictions; release is created by musical events that adhere more closely to the listener's predictions. The composer/performer combination engages listeners by providing enough surprises to sustain interest. Listeners will lose interest at the extremes—i.e., if the music either contains so many surprises that it overwhelms the listener's syntactical understanding, or becomes so predictable as to create boredom.[13]

Meyer's theory seems entirely consistent with another passage in Markind's chapter 1:

> As the brain listens to music, the auditory regions share predictions about upcoming sounds with the reward system. The RPE [reward] system is activated if there is a mismatch of the next sounds with what was predicted. This, in turn, generates sentiments.

Markind's prose makes the scientific bases for music clear and accessible to lay readers, which makes this book useful to parents, educators, policymakers . . . indeed, anyone who cares about children receiving a balanced education. The objective evidence presented here should convince any open-minded person of the power of music in education and in daily life, hopefully including those who have hitherto stubbornly refused to believe that anything as enjoyable as music could possibly have practical importance in children's development and education.

I encourage readers to take seriously Markind's suggestion that you involve yourselves actively in music-making. Not only will your experience of the joys of—and, yes, the discipline required by—music make the case for music education more eloquently than could any book, but it will kindle a powerful new light in your life.

Dr. Shuler is a Past President of the National Association for Music Education (NAfME). Experienced in curriculum development at state and national levels, he co-chaired the development of America's current National Standards in Music Education.

Appendix A

An Essential Brain Network for Music

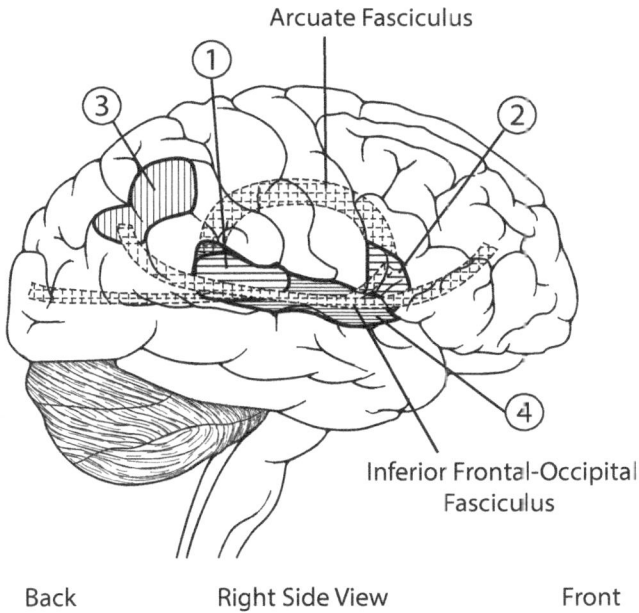

FIGURE A.1   Schematic of an essential brain network for music. Key: (1) posterior portion of superior temporal gyrus, (2) inferior frontal gyrus, (3) inferior parietal lobule, (4) anterior portion of superior temporal gyrus. The critical white matter tracts of the network—the arcuate fasciculus and the inferior frontal-occipital fasciculus—are shown as cross-hatching and labeled.

Adapted from Aleksi Sihvonen et al., "Neural Architectures of Music—Insights from Acquired Amusia," *Neuroscience & Biobehavioral Reviews* 107 (2019): 111, https://doi.org/10.1016/j.neubiorev.2019.08.023, with the kind permission of Aleksi Sihvonen.

This schematic of the right hemisphere shows a "dual-stream" model, the combined functions of the dorsal (upper) and ventral (lower) streams forming the music network proposed by Sihvonen et al.[1] The dorsal stream includes the posterior portion of the superior temporal gyrus (1), inferior frontal gyrus (2), and the arcuate fasciculus, which is the white matter tract that connects them. The ventral stream comprises the inferior parietal lobule (3), anterior portion of the superior temporal gyrus (4), inferior frontal gyrus (2), and the white matter tract connecting them known as the inferior frontal-occipital fasciculus (IFOF). Not labeled: The most forward aspect of the IFOF projects to the pre-frontal region, seat of working memory with "its capacity to maintain sounds in memory over time."[2] (The rearmost portion of the IFOF projects to the occipital lobe, center of vision, hence minimally involved in music function.)

Robert Zatorre summarizes the key functional role of the dorsal stream as relating to temporal (time-based) predictions of sound "especially important for the tracking of rhythms and for processing of the musical beat," while the key functional role of the ventral steam is "especially important for operations that enable us to understand relationships between sounds and how these form patterns."[3]

Aleksi Sihvonen's group reports that recovery from amusia is unlikely if both steams in the right hemisphere are damaged. However, recovery is possible if either one of the streams is preserved.[4]

---

1 Sihvonen, "Neural Architectures of Music."
2 Zatorre, "Musical Enjoyment and the Reward Circuits of the Brain," 439.
3 Zatorre, "Musical Enjoyment and the Reward Circuits of the Brain," 439–440. Zatorre provides a comprehensive discussion of the functional roles of each stream in Zatorre, *From Perception to Pleasure*, 65–135.
4 Sihvonen, "Neural Architectures of Music," 111.

Appendix B
Famous Neuroscientists Appearing in This Book

Alois Alzheimer (1864–1915)
In 1901, Alzheimer met Auguste Dieter, a woman hospitalized at a long-term facility. He observed and documented the symptoms and signs of her illness until her death, in 1906. When he reported her clinical illness and brain autopsy findings the following year, Alzheimer thought he was describing a rare condition. This remained the conventional wisdom until 1976, when Robert Katzman wrote "The Prevalence and Malignancy of Alzheimer's Disease," an editorial asserting that the condition—designated Alzheimer's disease in 1910—was, in fact, widespread among the elderly and would become increasingly prevalent due to the aging population. Time has proven Katzman's prescience.

Pierre Paul Broca (1824–1880)
Broca, a renowned physician and anatomist, will forever be associated with the region of the left frontal lobe that bears his name: Broca's area. He studied the aphasia (language problem) of Louis Victor Leborgne, whose illness left him with the ability to speak only the sound "tan." After Leborgne's death, Broca studied his brain and discovered a lesion of the left frontal lobe. Correlating that lesion with Leborgne's aphasia greatly accelerated acceptance of the concept of localization. See figure 2.1 for the location of Broca's area.

Jean-Martin Charcot (1825–1893)
Charcot was the most renowned neurologist in France during the late nineteenth century. He was also famous for his Socratic teaching method. Each Tuesday he would gather his students and assistants, called residents today, to examine and discuss

patients. The sessions were transcribed and collected as a series, Les Leçons du Mardi à la Salpêtrière (Tuesday Teaching Rounds at the Salpêtrière Hospital). Later in his career, Charcot's curiosity and interest turned to hysteria and hypnosis. Of note, one of his students was Sigmund Freud. Attracted to the evolving bohemian entertainment scene in Montmartre, Charcot introduced the young Freud to it after hours.

Antonio Damasio (1944–)

Damasio's clinical and laboratory work places him at the nexus of neurology and philosophy. That is, through scientific study of the brain he investigates what we can know and how we can know it. He is recognized internationally as a leader in studies of consciousness and of the self, the latter being something of a paradox because studying it scientifically requires objectively probing the subjective. Damasio has published his profound ventures into the brain and mind as a series of highly acclaimed books, beginning with the intriguingly titled *Descartes' Error* (1994).

Michael Gazzaniga (1939–)

Gazzaniga is a leading researcher in cognitive neuroscience. Of particular note, his study of "split brain" patients has yielded a trove of information about how the brain works, particularly how the cerebral hemispheres sometimes function dependently and sometimes independently of one another. His mentor, Roger Sperry (1913–94) won the 1981 Nobel Prize in Physiology and Medicine for pioneering this work; after Sperry's death, Gazzaniga became the recognized leader of the field.

Norman Geschwind (1926–1984)

Geschwind reportedly became intrigued by human behavior on the battlefield during his World War II military service. Subsequently planning to go into psychiatry, he was "bothered by the general lack of understanding of and interest in the basic neurology behind the psychology at the time."[1] Exposure to and study of numerous patients with language difficulties led him to describe the concept of disconnection syndrome, wherein the white matter connection between functional gray matter modules is interrupted, and to develop the nomenclature used to this day to describe various types of aphasia. He is credited with coining the term *behavioral neurology* and is therefore recognized as the "father of behavioral neurology" in the United States. He mentored numerous residents and students, many of whom became leaders in their fields. Naomi Geschwind, his daughter, recalls a rich intellectual atmosphere when her parents welcomed residents and students to their home.[2]

Donald Hebb (1904–1985)

Hebb, a psychologist interested in the physiology of the nervous system, studied the cellular basis of learning. His conclusions led to the coining of the expression "neurons that fire together, wire together" to describe how connections in the brain form and are reinforced. This phrase has entered popular awareness at many levels of the culture.

John Hughlings Jackson (1835–1911)

A British physician, Hughlings Jackson's valuable insights into brain function cannot be underestimated. He created the conceptual framework that underlies the discipline of clinical neurology. He was an advocate of localization at a time when many remained skeptical of the idea. Hughlings Jackson viewed the brain as an organ comprising multiple layers that developed

---

1 Geschwind, "Are You Related to 'the Geschwind?,'" 124.
2 Naomi Geschwind, personal conversations

over the course of evolution; positive and negative symptoms of brain disorders result from the interplay among these layers.[3]

Daniel Levitin (1957–)

Levitin worked as a musician, sound engineer, and record producer before becoming a neuroscientist studying music. This diverse set of experiences enables him to bridge the gap between the art and science of music with great skill. His book, *This is Your Brain on Music* (2006), is a classic in the field of popularizing the scientific study of music.

Lionel Naccache (1969–)

A behavioral neurologist in Paris, Naccache works with a team at the Brain Institute of the Salpêtrière Hospital exploring the neural bases of human cognitive functions. Naccache's research and publications in the French-speaking world complement Damasio's work in the Anglosphere. Naccache also co-authored a proof-of-concept study involving listening to classical music to improve post-stroke aphasia, the proposed mechanism being enhanced brain connectivity.[4]

Wilder Penfield (1891–1976)

Penfield helped found the Montreal Neurological Institute in 1934, after training under some of the greatest names in neuroscience. As the institute's first director, Penfield established the worldwide reputation of the Neuro (as it's called in local parlance), both by his own accomplishments as well as by recruiting top talent to his team. He is perhaps best known for *Epilepsy and the Functional Anatomy of the Human Brain* (1954), in which

---

3   For a superbly interesting short biography, see the introduction to "An Introduction to the Life and Work of John Hughlings Jackson."
4   Chea et al., "Listening to Classical Music Influences Brain Connectivity in Post-Stroke Aphasia."

he and coauthor Herbert Jasper described the brain's representation of the body known as the homunculus (see Figure 2.2).
Penfield obtained information that led to the discovery of the
homunculus by applying gentle electrical stimulation to the
brain during surgery on patients while they were awake.

Isabelle Peretz (1956–)

Peretz, a leading music neuroscientist, collaborated with Robert Zatorre to establish BRAMS, the International Laboratory
for Brain, Music, and Sound Research. The renowned organization greatly contributes to Montreal's reputation as an international center for the scientific study of music and auditory
cognition.

Hervé Platel (1965–)

Platel is a professor of neuropsychology at the University of
Caen in Normandy, France. His laboratory and clinical research
interests focus on the role of artistic practice in understanding
cerebral plasticity and its potential use in the management of
individuals with brain lesions. Of particular note, he envisions
the application of music as part of the management of patients
with Alzheimer's dementia.[5]

Oliver Sacks (1933–2015)

Quite simply, the greatest author the field of neurology has ever
produced, Sacks's fame derived from his interest in closely
following patients—often for extended periods of time—and
his ability to describe his observations in prose so accessible
that the general public readily grasped and enjoyed reading it.
With respect to music and the brain, *Musicophilia* (2007) is a
classic and stands as his outstanding contribution. Sacks was

---

5   https://editions-attribut.com/portfolio/herve-platel/

instrumental in establishing the Institute for Music and Neurologic Function, currently located in Mount Vernon, New York, and led by Concetta Tomaino, a renowned music therapist who worked with Sacks for many years.

Michael Thaut (1952–)

Thaut is a founder of the field of neurological music therapy and co-author of the acclaimed textbook *An Introduction to Music Therapy: Theory and Practice*. He is also an international research leader in the neuroscience of music and the applications of auditory neuroscience to neurological rehabilitation.[6]

Carl Wernicke (1848–1905)

Wernicke published a paper in 1874 about a type of aphasia different from the kind described by Broca. These patients could not properly comprehend speech and, while they could speak fluently, what came out of their mouth was generally gibberish since they were not able to understand even their own speech. This condition is called a receptive aphasia whereas Broca's aphasia is called an expressive aphasia. Brain autopsy demonstrated a lesion in the posterior temporal lobe, now known as Wernicke's area (see Figure 2.1 for the location of Wernicke's area).

Robert Zatorre (1955–)

Zatorre, a leading music neuroscientist, collaborated with Isabelle Peretz to establish BRAMS, the International Laboratory for Brain, Music, and Sound Research. The renowned organization greatly contributes to Montreal's reputation as an international center for the scientific study of music and auditory cognition.

---

6   For more information, see "Michael Taut," https://rsi.utoronto.ca/faculty /michael-thaut.

Glossary

Acquired: A condition that a person is not born with, as opposed to a congenital condition.

Amusia: While the term *amusia* literally means "without music," it is generally used by neurologists and neuroscientists to denote any abnormality of the brain's musical functioning. In most cases, the aberration is a decline, or under expression, of the brain's musical functioning; occasionally, an excess, or over expression, of musical functioning is observed. Amusia often involves an identifiable lesion of the brain. This is important scientifically because lesion studies are essential in identifying brain regions causally linked— that is, having a verifiable cause and effect relationship—to a given capacity or skill.

Aphasia: Any abnormality of the brain's language functioning.

Aprosody: An inability either to produce or to comprehend the affective (sentimental) components of speech.

Cerebrovascular support: The richness of the blood supply to the brain. The majority of strokes occur when blood supply to an area of the brain becomes blocked, so better cerebrovascular support confers a reduced risk of stroke.

Cingulate gyrus: A deep structure in each hemisphere of the brain, different types of memory are associated with its various parts. Although it looks like a single, belt-shaped (from the Latin *cingulatus*) unit, the anterior cingulate gyrus belongs to the limbic system, while the posterior cingulate gyrus belongs to the default network.

Cognitive reserve: The ability of the brain to access or process information through extra or redundant cerebral pathways. Loss of cerebral pathways results from brain disease and is also a feature of normal aging. A brain with more pathways has greater cognitive reserve and is thus less vulnerable to aging or brain disease.

Congenital: Refers to a condition present at birth, as opposed to an acquired condition. The most common congenital disorder of music is tone deafness, found in about 3 percent of the population. Since it is an abnormality of music function from birth, tone deafness is an example of a congenital amusia.

Corpus callosum: A white-matter structure that serves as the principal information superhighway connecting the two hemispheres of the brain.

Expressive: Indicating involvement of one's ability to produce something. For example, an expressive **amusia** indicates difficulty producing one or more aspects of music. The inverse of expressive is **receptive.**

Gray matter: Areas of the brain where cell bodies, the part of the cell where the nucleus resides, of neurons are present.

Gyrus (plural *gyri*): The folds or bulges of the brain.

Hemiparesis: Weakness on one side of the body.

Hertz (Hz): The cycles per second of a sound wave; named for Heinrich Rudolf Hertz. Musical notes are identified by their frequency as measured in Hertz. Figure 4.1 shows one cycle of each sound wave pictured. Musical notes differ from one another by virtue of their frequencies. For example, the note A above middle C—the note to which an orchestra tunes—is defined as 440 Hz. Because this A is located in the fourth octave of a piano, it can be notated as A_4. Its frequency is twice as fast as A_3, four times as fast as A_2, half as fast as A_5, and so on. The frequency of a note is twice as fast as the same note in the octave just below it. There are twelve notes in each octave—the interval between two notes with the same name—and there is equal frequency "spacing" between any two notes.[1] This spacing can be mathematically expressed as a

---

1   The technical term for this is *equal temperament*. It's a very practical way to relate notes to one another, but it's not the only way. For a concise discussion of this and other tuning systems, the interested reader may wish to consult Sulzer, *Music, Math, and Mind*, 33–62.

multiple, a number multiplied to the previous note. Since an octave is defined as a multiple of two (the frequency of A_4 is twice that of A_3), it follows that the note just above A_4, called A-sharp and designated as $A^\#_4$, is 1.06 times the frequency of A_4, or 466.4 Hz.[2]

Hippocampus: Each hemisphere of the brain has a hippocampus in the temporal lobe. The hippocampus is necessary to form new autobiographic memories, which was proven when the hippocampi of Henry Molaison were surgically removed in 1953. A similar fate—the loss of the ability to form new autobiographic memories—befell a musician named Clive Wearing after he lost the functioning of both hippocampi due to an infection in 1985.

Incidence: How many new cases of a condition are recognized per unit time, usually calculated on a per year basis.

Index case: The first documented case of a condition. For example, Louis-Victor Leborgne is the index case of Broca's aphasia, and Auguste Dieter is the index case of Alzheimer's disease.

Infant directed speech: A manner of grammatically correct speech directed at infants and characterized by warm timbres, slow tempos, elevated pitches, and expressive dynamics (variations of volume); is also called "motherese" or "parentese."

Interference: The interaction of two sound waves with each other. Interference can add to one another (constructive) or subtract from one another (destructive). This is graphically represented and described in figure 4.1.

Lesion: A general, nonspecific term referring to some sort of objective abnormality of a body part or on an imaging study.

Lesion method: Investigating what happens when a part of the brain malfunctions or is injured. Were the brain an external circuit board, the lesion method would be akin to removing a circuit and studying what occurs as the remaining circuits go about their

---

2 It's 1.06 because that is the twelfth root of 2; expressed another way, multiplying 1.06 times itself 12 times equals 2.

normal business. However, the brain is not external to our bodies and we cannot go around removing its circuits. Instead, what we are able to do is to observe the impact upon abilities in people who, for whatever reason, happen to have impaired functioning (due to a lesion) in a particular brain region.

Localization: The concept that different parts of the brain are genetically destined to perform different functions. It also refers to the observation that brain operations are broken down into components that occur in spatially separated areas of the brain.

Neurons: Brain cells that transmit information and signal one another.

Neuroplasticity: The brain's ability to adapt to novel information or circumstances by forming new neural circuits or revising established ones. A more plastic brain forms these circuits faster and more robustly. On a cellular level, neuroplasticity refers to the number of neuronal connections or the rapidity with which these connections are made. Since the brain can update itself through exposure to different experiences, the more experiences a brain has, the more connections it makes.

Neurotransmitter: A chemical released by a cell in the nervous system that carries information to one or more other cells within the system.

Plasticity: See **Neuroplasticity.**

Prevalence: The total number of cases of a condition in a population at a given time.

Proprioception: Sensing a body part's motion and location in space.

Prosody: Patterns of stress and intonation, such as voice inflection, that impart sentimental features to spoken language; it is sometimes referred to as the musical qualities of speech.

Receptive: Involving one's ability to perceive something. For example, a receptive amusia indicates difficulty with perception of one or more aspects of music. The inverse of receptive is expressive.

Somatotopic: Means that the brain, by sorting the nerve signals it receives according to their location of origin in the body, creates a map of those signals. See figure 2.2 for example.

Spectral discrimination: The capacity to distinguish sound wave frequencies.

Stroke: The blockage of a blood vessel, preventing blood from reaching a part of the brain, or the hemorrhaging of blood into the brain, both causing that part of the brain to die as a result. The former type is an ischemic stroke, and the latter is a hemorrhagic stroke, which is the less frequent of the two. To learn more about stroke facts, prevention, and treatment, visit the website of the American Stroke Association, https://www.stroke.org/.

Sulcus (plural, *sulci*): The sulci are the indentations or grooves of the brain.

Tone deafness: Difficulty perceiving or producing small changes in pitch. Humans can typically detect or produce a pitch change of one-half of a semitone (a semitone may also be called a "half-step"). Those who are tone deaf are unable to perceive or produce a pitch change of less than two semitones.

Tonotopic: The cochlea and brain's ability to sort the auditory nerve signals they receive according to the signals' wavelengths, thus creating a map of those sound signals. See figure 2.4.

Tractography: A neuroimaging technique used to demonstrate the functioning of white matter pathways (tracts) of the brain. Functional tractography exposes the presence of physiologic activity—as opposed to displaying the anatomy—of a white matter tract being studied.

Vocal learning: The ability to imitate and acquire novel sounds. Humans, songbirds, and certain marine animals (such as members of the whale and seal families) have this ability. Curiously, nonhuman primates (such as chimpanzees, gorillas, and even singing gibbons) do not.

White matter: Areas of the brain that do not contain cell bodies. The white color indicates the presence of myelin, an insulating material around nerve projections called axons. Myelin serves the same function as insulation around a copper wire, greatly speeding the transmission of the electrical signal coursing through the wire.

Acknowledgments

I wish to thank a number of people whose contributions helped make this book a reality:

Amy, Brad, and Gary—my "classmates" from a course for first-time authors, for helping one another climb the mountain; Bailey, Brandi, and Jenn—our teachers, for sharing their wisdom and guiding us, keeping us on track

Dwyer Conklin and Hervé Platel for generously sharing their insights about the present and future of their fields—music therapy and music neuroscience, respectively—that grace the pages of this book

Joanna Bereaud for providing music therapy to the youngest among us

Cynthia Bahr for offering music to help patients with movement disorders

Stacie Lee Martin Giles, Evlyn Gould, Steven Herrine, Susan Knoppow, Dina Markind, Jan Mashman, Paul Nickolloff, and Jeff Thatcher for taking the time to review and critique chapters of this book

Stacie Lee Martin Giles for reconnecting after more than forty years and providing critical help and guidance, to you extra thanks

La famille Ruimy: Je vous embrasse très fort

Jennie Riblet for accepting my gratitude on behalf of her mother, Linda

Allan and Vikki for sharing your journeys with me and the readers of this book, many thanks

Sheilah Rae and Scott Shuler for kindly sharing their thoughts in the foreword and afterword, respectively

Allan Graubard, independent editor and agent, and Suzanne Staszak-Silva, senior acquisitions editor at Johns Hopkins University Press, for believing in this book; also at JHUP, Charles Dibble, production editor and fellow Charles Aznavour fan, and Marlee Brooks, editorial assistant and like me a graduate of Auburn (NY) High School, for guiding this book through copyediting, proofreading, and printing

Lou and Andy at Arkettype for their outstanding graphics support

Dina for our wonderful life together. I couldn't have done this without you.

Notes

Preface

1. Panter, *Creativity & Madness*, xi. For information on activities and conferences, visit Creativity and Madness, https://www.creativityandmadness.com/.

2. The Institute for Music and Neurologic Function (https://www.imnf.org/) was originally located at the Beth Abraham Home in the Bronx. It is now based in Mount Vernon, New York.

Introduction

1. Pinker, *How the Mind Works*, 534: "I suspect that music is auditory cheesecake, an exquisite confection crafted to tickle the sensitive spots of . . . our mental faculties." Pinker also wrote, "Music could vanish from our species and the rest of our lifestyle would be virtually unchanged" (528). In the foreword to the 2009 edition of *How the Mind Works*, Pinker humorously acknowledges the impact the term *auditory cheesecake* has had, noting that it "will probably appear in my obituary."

2. Spitzer, *The Musical Human*, 5.

3. EMI (Electric & Musical Industries), best known for producing and selling Beatles records, also developed electronic equipment. Godfrey Hounsfield, an engineer working at EMI, was one of the principal creators of CT technology, for which he shared the 1979 Nobel Prize in Physiology or Medicine. See Goodrich, "The CT Scanner Was Invented by a Music Company Engineer." On a personal note, a man originally from Britain walked into my office one day around 2015 for a neurological consultation. He had also worked at EMI, on a project to develop magnetic resonance (MR) imaging. Pulling a picture of the scan from his pocket as proof, he told me that his brain was the first to be imaged using MR technology, a landmark accomplishment that took place at an EMI laboratory in 1978.

4. Peretz and Zatorre, *The Cognitive Neuroscience of Music*. In December 2023, the National Institutes of Health sponsored "Music as Medicine: The Science and Clinical Practice," a hybrid (in-person and online) symposium that reached a broad audience. The keynote speaker was the opera soprano Renée Fleming. Much of the material presented at the symposium appears in Fleming, *Music and Mind*.

5. The term *survival value* can encompass both survival and procreation, the former referring to survival of an individual, and the latter to survival of the species.

6. From this neurologist's perspective, "music will always be of the neocortex and of the limbic system" would be a more neurologically accurate way to express this. In reality, music is all about the brain, but there is nothing charming, poetic, or memorable about such phrasing.

1. The Basics of a Musical Brain

1. Wurz, "Interpreting the Fossil Evidence for the Evolutionary Origins of Music," 411–12.

2. For images of the flutes, see "Eight Oldest Musical Instruments in the World."

3. The Divje Babe flute, found in Slovenia and made of bear thigh [femur] bone, has been dated to 50,000–60,000 years ago. It may be of Neanderthal origin; however, it is disputed whether this was truly a musical instrument or just a piece of bone chewed on by a scavenger animal. See Martins, "Was 'Earliest Musical Instrument' Just a Chewed-Up Bone?"; and "Neanderthal Flute: The Oldest Musical Instrument in the World."

4. To watch the Hohle Fels flute being played, see Service, "The Ice Age Flute That Can Play the Star-Spangled Banner."

5. Bones resist decomposition far better than skins and wooden frames, so no remnant of a membranophone older than 5,000 or 6,000 years old exists any longer, although such an instrument may have been invented earlier.

6. See Svard, "Music and Prehistoric Cave Art." For this reference, I thank Dr. Jocelyn Soffer, a child and adolescent psychiatrist in private practice who also loves to sing and designed and teaches the course Singin' in the Brain: Music and the Developing Child at New York University.

7. *Tanakh: A New Translation of the Holy Scriptures According to the Traditional Hebrew Text*, 1224.

8. Music along with arithmetic, astronomy, and geometry comprised the quadrivium, the four sciences, dating back to the writings of Plato. Thaut, "Coda and Crescendo," 421.

9. Watson, "The Songs of Gods and Men," 28.

10. Watson, 27.

11. Ephesians 5:19, in *The Holy Bible,* Revised Standard Version, second ed., ed. Jeff Cavins et al. (Oxford University Press. 1963), 183.

12. DeVale, "Harp: IV. Asia," 2001.

13. Ragahavan quoted in Beck, *Sonic Theology*, 107–8.

14. Davies, *Ancient Egyptian Paintings*, 94–97.

15. See image at Wikimedia.org, https://upload.wikimedia.org/wikipedia /commons/thumb/6/6e/Tomb_of_Nakht_-_three_musicians_600dpi.png /1200px-Tomb_of_Nakht_-_three_musicians_600dpi.png.

16. "Eight Oldest Musical Instruments in the World."

17. See image at Wikipedia.org, https://en.m.wikipedia.org/wiki/File: Silver_trumpet_from_Tutankhamun%27s_tomb.jpg.

18. Copland, *What to Listen for in Music,* 34–35. While the fossil record reveals bone flutes to be older than membranophones, simple rocks banged together (lithophones) would not come to anyone's attention.

19. Rogosin quoted in Panter, *Creativity & Madness,* 3:200.

20. From Oliver Sack's statement at "Forever Young: Music and Aging," a hearing before the Special Committee on Aging of the United States Senate, August 1, 1991, quoted in Tomaino, "Music Has Power® in Senior Wellness and Healthcare," 217.

21. The aspect of speed, formally called tempo, can play a role in imparting meaning in speech in the sense that saying something slowly or quickly can change its meaning. Tempo does not, however, form a constraint around speech. Consider the difference between driving a car on a busy freeway versus in an empty parking lot. On a freeway, you must (are constrained to) keep moving in the direction of traffic, maintain your speed within a certain range, and stay within the lines of your lane. By contrast, you are at liberty in the empty parking lot to travel in any direction you choose, to alter your speed to whatever you wish, and to cross over the painted lines with impunity. Music is like the freeway. Rhythm compels the music forward at a sustained pace with a continuous beat. Speech is akin to the empty parking lot. You can go at whatever speed you want—you can even stop for a break and then continue—and choose the direction that suits you. The closest speech comes to music rhythmically is through poetry. No surprise here: Lyrics are poetry set to music.

22. Matching bodily movement to an external rhythm is called entrainment, a subject discussed at length in chapter 4.

23. Copland, *What to Listen for in Music*, 49.

24. Copland, 51. Contrasting the "musicality" of music and speech, music neuroscientist Aniruddh Patel wrote, "If a musical melody is 'a group of tones in love with each other' then a linguistic melody is a group of tones that work together to get a job done." Patel, *Music, Language, and the Brain*, 184. Patel attributes the phrase "a group of tones in love with each other" to the composer and oud virtuoso Simon Shaheen.

25. A pitch is expressed in the world of music as a musical note. I use the words *tone, pitch,* and *note* interchangeably.

26. Copland, *What to Listen for in Music*, 60.

27. Bennett Lerner, personal communication, 2016. Lerner was a young pianist who knew Copland well at the end of Copland's life. Lerner reported to me that Bernstein recognized Copland's decline when "he felt Copland's usually tight-yet-expansive long line was no longer fully evident."

28. Copland, *What to Listen for in Music*, 50.

29. As harmony developed, it gradually assumed the principal role of defining the style of a piece of music. That had previously been the role of modes, which are still used today. As harmony became increasingly complex, however, modes became subsumed into the overall category of harmony.

A composer constructs a song using certain notes the way an artist creates a piece using certain materials. The notes in the composer's "quiver" form a scale. There are twelve notes in each octave (ignoring that the word *octave* derives from a Latin root meaning "eight"); this is called a chromatic scale. The distance between any two adjacent notes is called a half step. Scales used in the construction of a song most commonly contain seven of the twelve notes within each octave (a diatonic scale), although some pieces of music use only five of the twelve notes within each octave (a pentatonic scale). This is accomplished by omitting certain pitches of the chromatic scale: sometimes utilizing a whole step between notes in a scale (a whole step meaning two half steps) and sometimes a step and a half. For example, the distance between C and C# is a half step—the terms *half step* and *semitone* are synonymous—and the distance between C# and D is a half step; the distance between C and D is therefore a whole step. The C-major diatonic scale—C-D-E-F-G-A-B—includes C and D but skips C#. It's the combination of steps sizes (half, whole, one and a half) within an octave that defines a scale.

The modes, or modal scales, of classical Greece were defined by each possible seven-note combination of half steps and whole steps. Of course, no audio exists of how a piece of music sounded in ancient Greece, but written records document that seven different modal scales were utilized. Western music today leans most heavily on two of these modes—called Ionian and Aeolian in ancient Greece and major and minor, respectively today. Other modes, that is, the five Greek modes beyond Ionian and Aeolian as well as non-Greek modes, are retained to this day in jazz and in non-Western musical traditions, for example, in eastern European–based "ethnic" music such as klezmer. Each mode conveys its unique character to a piece of music, which is likely why many different modes have been used over time.

30. For example, *A Fifth of Beethoven*, a 1976 jazzy disco arrangement by Walter Murphy adopts the famous melodic line from the first movement of Beethoven's Fifth Symphony.

31. Copland, *What to Listen for in Music*, 60–61, calls harmony one of "the most original conceptions of the human mind."

32. Copland, 62.

33. Copland, 78.

34. The profile of overtones provides the spectral (sound wave frequencies) component of timbre, but timbre also has a temporal component. This was demonstrated in the 1950s by Pierre Schaeffer in his "cut bell" experiments. For more information, see Li, "How We Process Timbre (Tone)." Briefly, each sound source has a unique signature of how quickly its sound onsets (the attack), how long the sound remains stable (the sustain), and how rapidly the sound resolves (the fade). The temporal profile of timbre also contributes to recognizing the source of the sound.

35. For a detailed and accessible deeper dive into this subject, see Dubuc et al., "'Dusting Off the Triune Brain and the Limbic System."

36. These designations are used here for ease of comprehension. They are not rigid boxes. In fact, proponents of these positions engage in a good deal of discussion.

37. Sound signals travel from the ear to the brain's primary auditory cortex in about 20 to 40 milliseconds (msec, thousandths of a second) whereas vision signals take around 75 to 100 msec to travel from the eye to the brain's primary visual cortex. These figures were discovered decades ago through evoked potential testing. Daly and Pedley, *Current Practice of Clinical Electroencephalography*, 598 (visual evoked potentials) and 658 (auditory evoked potentials). See also Kraus, *Of Sound Mind*, 44.

38. "The ability to tap to the beat is unique to music . . . and is a natural behavior even in people with no musical training." Zatorre, Chen, and Penhume, "When the Brain Plays Music," 550.

39. "If an individual stands completely still while listening to music, functional magnetic resonance imaging (fMRI) scanning will show that his motor cortex is activated, although the individual is not in motion." Perrone-Capano, Volpicelli, and di Porzio, "Biological Bases of Human Musicality," 237.

40. Konoike and Nakamura, "Cerebral Substrates for Controlling Rhythmic Movements," 1.

41. How the brain constructs a mental concept of time is not clearly understood, in part because time is a sense that has no specific receptors as opposed to hearing (the ears) or vision (the eyes). See Konoike and Nakamura, "Cerebral Substrates for Controlling Rhythmic Movements," 7.

42. Zatorre, Chen, and Penhume, "When the Brain Plays Music," 548.

43. Other sensations involved include proprioception and time. Konoike and Nakamura, "Cerebral Substrates for Controlling Rhythmic Movements," 6, report asymmetric (right greater than left) activation of the parietal lobe with respect to representation of the sense of time.

44. Sankaran, Leonard, Theunissen, and Chang, "Encoding of Melody in the Human Auditory Cortex," 1.

45. Less than 5 percent of the population has absolute pitch, also known as perfect pitch, and the actual number may be less than 1 percent, depending on how the term is defined and the data are collected. See Witynski, "Perfect Pitch, Explained."

46. Not needing to refer to a fixed starting note explains how transposition is possible. As long as the intervals between the notes are preserved, the brain will identify the melody as the same regardless of which note it starts on (i.e., which key it's in). See Adachi and Trehub, "Musical Lives of Infants," 231–32, for details on the ability of newborns to perceive pitch relations.

47. In humans, this function is specialized for music. Sankaran, Leonard, Theunissen, and Chang, "Encoding of Melody in the Human Auditory Cortex," 9.

48. In the septal area per Olds and Milner, "Positive Reinforcement Produced by Electrical Stimulation of Septal Area and Other Regions of the Rat Brain," 419; specifically in the nucleus accumbens per Peretz, *How Music Sculpts Our Brain*, 6.

49. "[A]n organism can more effectively prepare an appropriate response to an event if that event can be predicted." Zatorre, "Why Do We Love Music?" 4. Therefore, predicting events has survival, hence evolutionary, value.

50. Martínez-Molina et al., "Neural Correlates of Specific Musical Anhedonia," E7337.

51. The portrait was a copy, the original having been destroyed by fire in 1698. "Portrait of Henry VIII," Wikipedia, https://en.wikipedia.org/wiki/Portrait_of_Henry_VIII.

52. Chapter 2 looks at a key brain region in the pathway for stimulating a MEAM. It differs from a nearby area—which is impaired by the most common form of dementia—that the brain uses when voluntarily trying to re-

member an event from the past. These, along with the therapeutic leveraging of music memory's distinctive qualities, are topics discussed in chapter 5.

53. See, respectively, Platel et al., "Semantic and Episodic Memory of Music Are Subserved by Distinct Neural Networks"; Platel and Groussard, "Benefits and Limits of Musical Interventions in Pathological Aging."

54. Navarro, Martinón-Torres, and Salas, "Sensogenomics and the Biological Background Underlying Musical Stimuli: Perspectives for a New Era of Musical Research," 1454.

One study cited in this paper analyzed the expression of several immune response-related genes and showed that recreational music making has the ability to modulate the human stress response. In another study, investigating the impact on the audience of Mozart's Violin Concerto No. 3, microRNAs that influence dopamine metabolism and prevent neurodegeneration were found. Acknowledging that there is currently little research on this subject, the authors advocate the development of a new discipline to investigate the impact of sensorial input (hearing in the case of music) on gene expression.

Perrone-Capano, Volpicelli, and di Porzio, "Biological Bases of Human Musicality," 2, cite a report that birds and humans both express a gene for the protein FOXP2 (forkhead box protein P2). Mutations of this gene in humans are related to childhood impairments in pronouncing sounds while ablation of this gene in birds prevents song learning.

55. Allan Graubard notes that a single tone in isolation is referential—to silence. Of course, we can also hear something in deep silence: our breathing.

56. Radical modernists who champion works lacking rhythmic structure are off in a corner somewhere with other members of their small cadre, not communicating musically in a manner that is shared with the general public.

57. The minimalist viewpoint is discussed in detail in chapter 3, and a third approach can be found in chapter 5.

58. Gault, "Touch as a Substitute for Hearing in the Interpretation and Control of Speech," 122.

59. The sensory receptor sensory cells are Pacinian, Meissner's, Y, and Z. Gault, "Touch as a Substitute for Hearing in the Interpretation and Control of Speech," 123, does not name them, but refers to leveraging the vibration range characteristics of these receptors for the appreciation of music.

60. This is an impressive range of four and a half octaves

2. A Closer Look at a Musical Brain

1. Collins, "Does the Bible Contradict Accepted Biological Concepts?"

2. Brandt, "Brain Beats Heart," 1617.

3. Pandya, "Understanding Brain, Mind and Soul," 131.

4. Pearce, "Rufus of Ephesus."

5. His description included a loop of arteries at the base of the brain that to this day is called the Circle of Willis, in honor of his work.

6. This might seem readily apparent, but the concept was passively avoided and actively shunned for generations. Humans tend to resist thinking of themselves as yet another member of the animal kingdom, however, a full understanding of the brain requires that we do precisely that, especially because the human brain includes layers that evolved long before our appearance on earth.

7. Dennett, "Review of Damasio, Descartes' Error," 3–4.

8. Proposed in the nineteenth century, this keen observation can be attributed to the great English physician John Hughlings Jackson. See appendix B for information on him and other famous neuroscientists mentioned in this book.

9. This point may seem obvious, yet is often overlooked when people—from natural philosophers to everyday folks—pose the question Why do I feel so connected to my body? This issue, called binding, need not be vexing if one hews closely to the axiom that brains serve bodies. The brain, being of the body, senses messages about the body as being in the body. Curiously, the brain remains oblivious to the fact that perception occurs within the brain itself.

10. For clarity in this book, I generally associate the middle tier with the limbic brain, emphasizing the role the limbic system plays with music. Nevertheless, there is a close connection between motor and behavioral skills, as the resemblance of the words motion and emotion attests. This connection is explored in detail in chapter 4.

11. Allen, *The Lives of the Brain*, 179. By comparison, brain energy consumption for chimpanzees, humans' closest living relative, is about 13 percent.

12. Making targeted lesions in animal brains has been done, but the information obtainable from animals is limited and not necessarily applicable to humans. Techniques to induce temporary functional lesions in the human brain are being developed and honed, such as transcranial direct current stimulation (tDCS). An application of tDCS is discussed later in this chapter (see note 40 below).

13. The term *somatosensory* refers to sensations that can occur anywhere in the body (soma). Examples include sensing vibration or the position of a

body part. Penfield's monumental work can be found in Penfield and Jasper, *Epilepsy and the Functional Anatomy of the Human Brain.*

14. Of note, the brain has no representation or map cf itself. As a result, it cannot be aware of itself directly.

15. Gazzaniga, *Who's in Charge?*, 19.

16. Hebb, *The Organization of Behavior*, 62: "When an axon of cell A is near enough to excite a cell B and repeatedly or persistently takes part in firing it, some growth process or metabolic change takes place in one or both cells such that A's efficiency, as one of the cells firing B, is increased."

17. Gazzaniga, *Who's in Charge?*, 13.

18. Kraus, *Of Sound Mind*, 40–43. See also Perrone-Capano, Volpicelli, and di Porzio, "Biological Bases of Human Musicality," 6.

19. Zatorre, Chen, and Penhume, "When the Brain Plays Music," 550.

20. See chapter 1, note 39.

21. Konoike and Nakamura, "Cerebral Substrates for Controlling Rhythmic Movements," 6.

22. Penhune and Zatorre, "Rhythm and Time in the Premotor Cortex," 1.

23. "Premotor Cortex."

24. Konoike and Nakamura, "Cerebral Substrates for Controlling Rhythmic Movements," 6.

25. Pointing out the greater importance of the right hemisphere versus the left hemisphere in this function, Battelli et al., "The 'When' Parietal Pathway Explored by Lesion Studies," wrote, "[The] right parietal lobe, and in particular the IPL, might play an important role in discriminating events that are displaced in time" (22).

26. A group of scientists—Konoike et al., "Temporal and Motor Representation of Rhythm in Fronto-Parietal Cortical Areas"—examined rhythm in two distinct ways: as a series of (incoming) sounds and as a series of planned (outgoing) movements. Their study results showed that when rhythm was an analysis of incoming sounds, the brain's two hemispheres activated symmetrically. When rhythm was studied as a sequence of potential movements—even if someone is lying still while imagining the motions—brain activity was asymmetric, greater on the right hemisphere. They deduced from this that the sequencing of rhythm to enable the transition from incoming sounds to outgoing movements is stored in the right hemisphere.

27. Konoike and Nakamura, "Cerebral Substrates for Controlling Rhythmic Movements," 4; Zatorre, Chen, and Penhume, "When the Brain Plays Music," 553.

28. Konoike and Nakamura. "Cerebral Substrates for Controlling Rhythmic Movements," 3; Kung, Chen, Zatorre, and Penhune, "Interacting Cortical and Basal Ganglia Networks Underlying Finding and Tapping to the Musical Beat," 417.

29. One other structure of note that plays a role in the assessment of rhythm is the labyrinth, the organ of the vestibular system. While it communicates with the brain, it is not part of the brain. An important sensor of movement, it lies adjacent to the cochlea in the inner ear. The labyrinth's role will be examined in detail in chapter 4

30. Nissim et al., "Frontal Structural Neural Correlates of Working Memory Performance in Older Adults," 1–2. The term *prefrontal* refers to the regions of the frontal lobes that are anterior to, that is, in front of, the frontal lobes' motor regions.

31. Rules of syntax are called *grammar* when applied to language.

32. Think hardware and software.

33. The dominant hemisphere is the hemisphere in which language function resides. It's often simply labeled as the left hemisphere since this is the dominant hemisphere for the vast majority of people.

34. Perrone-Capano, Volpicelli, and di Porzio, "Biological Bases of Human Musicality," 3–4.

35. Sankaran et al., "Encoding of Melody in the Human Auditory Cortex," 4: "The representation of melodic features did not strongly segregate into anatomically distinct subregions.".

36. Sankaran et al., 1: "Sites that encoded [absolute] pitch and [relative] pitch-change in music used the same neural code to represent equivalent properties of speech."

37. Zatorre, "Why Do We Love Music?," 4: "This [anticipation] is an essential ability for survival because an organism can more effectively prepare an appropriate response to an event if that event can be predicted."

38. Only minor differences of localization or lateralization of this activity were noted in Sankaran et al., "Encoding of Melody in the Human Auditory Cortex," 4–5: "In the right hemisphere, [relative] pitch-change was encoded anterior to expectation [i.e., next pitch prediction]," whereas, "the music selectivity [next-pitch prediction] was . . . weakly biased toward posterior regions in the left hemisphere."

39. Sankaran et al., 6: "The encoding of melodic expectation is functionally specialized for music."

40. Schaal et al., "Right Parietal Cortex Mediates Recognition Memory for Melodies," 1660. This conclusion results from data obtained from brain

imaging as well as from tDCS techniques, a noninvasive way to temporarily up- or down-regulate a selected region of the neocortex. It can therefore be used as a tool to leverage the lesion method by creating a transient "lesion" of the brain without entering the skull or causing irreversible damage.

41. Some authors even speak of "metaplasticity," a superior ability to acquire new brain skills in people who have had prior musical training. Vlaicu and Bustuchina, "La musique, un bon outil pour étudier le cerveau," 29.

42. Kandel, *The Age of Insight*, 36–47. On a historical note, Freud was a product of this school of thought and correctly recognized the fundamental concept—that much of what the brain does is hidden from consciousness. Certain details he championed, however, such as the Oedipus complex, failed to withstand the test of time and were discarded.

43. In figure 2.2, note that these are two slightly different maps situated close to one another: one is a map of the primary motor area of the neocortex and the other is a map of the primary somatosensory area of the neocortex.

44. Damasio, *Self Comes to Mind*, 199.

45. Lionel Naccache points out that we are not able to experience odors volitionally in our imagination because there is no primary olfactory cortex. Instead, olfactory information is routed directly to zones of the brain devoted to memory, emotional valence, and consciousness. Seizures that originate in these brain areas can induce smells as one of their symptoms, an example of involuntarily experiencing odors. Naccache, *Apologie de la discrétion*, 55–56.

46. Such as the reticular activating system, a collection of cells in the brain stem.

47. Humans cannot willingly be conscious of the activity in their association zones, but recombining information in novel ways by subconscious multimodal processing is an important aspect of creative thought. People may unexpectedly become aware of it in the form of dreams or sudden "inspiration."

48. Kahneman, *Thinking, Fast and Slow.*

49. Damasio, *Descartes' Error*, 128.

50. Damasio, 33.

51. The orbitofrontal region richly connects with the limbic brain and is "the key [neocortical] brain area in emotion, and in the representation of reward value." Rolls, Cheng, and Feng, "The Orbitofrontal Cortex," 1.

52. Signals can be sent to any of the body's organs and from there come to the conscious brain's awareness. I specify the gut in this example to leverage the expression "on a gut level."

53. Imagining beyond the near future is progressively less reliable, hence less accurate. Of note, significant overlap exists between memory of the past and predicting or imagining the future. Schacter et al., "The Future of Memory," 677. This comes as no surprise given that past experience contributes significantly to one's predictions about the future.

54. The learned values operate on a subconscious level once established, though they may have been initially acquired through conscious effort, such as studying or practicing them.

55. The part of the striatum called the nucleus accumbens is particularly active in this process, with greater involvement of the right nucleus accumbens as compared to the left nucleus accumbens. Gold et al., "Musical Reward Prediction Errors Engage the Nucleus Accumbens and Motivate Learning," 3311.

56. Martínez-Molina et al., "White Matter Microstructure Reflects Individual Differences in Music Reward Sensitivity," 5018: "[I]ndividual differences in music reward sensitivity are driven by variability in functional connectivity between the nucleus accumbens, a key structure of the reward system, and the right superior temporal gyrus.".

57. Zatorre, "Why Do We Love Music?," 8–9.

58. Approach may be observed along multiple axes, such as wanting (motivation), liking (hedonic), and learning (predictions). Ferreri et al., "Dopamine Modulates the Reward Experiences Elicited by Music," 3795.

59. In the scientific literature, such a response is generally called "chills." Damasio, in *Looking for Spinoza,* 102–3, sketches brain-mechanistic overview of chills.

60. Schlund and Cataldo, "Amygdala Involvement in Human Avoidance, Escape and Approach Behavior," 771.

61. The researchers wrote that this patient is also "severely limited in recognizing facial expressions of fear, whereas she is normal at recognizing happiness . . . in faces." Gosselin et al., "Amygdala Damage Impairs Emotion Recognition from Music," 237. While accepting that the amygdala's role is complex and "likely extends beyond aversive stimuli and fear," Inman et al., "Human Amygdala Stimulation Effects on Emotion Physiology and Emotional Experience," 2, reported, "Amygdala lesions have been observed to blunt normal physiological and subjective emotional responses to some types of emotionally salient stimuli, particularly fear-related responses."

62. Gosselin et al., "Amygdala Damage Impairs Emotion Recognition from Music," 240.

63. *Double dissociation* is the technical term for this.

64. Naccache, *Quatre exercices de pensée juive pour cerveaux réfléchis,* 44.

65. These findings were also demonstrated with Clive Wearing, a musician who suffered permanent bilateral hippocampal dysfunction following an infection (herpetic encephalitis) in 1985.

66. Dissociation between performance and awareness implies that Kahneman's system 2 thinking has access to only a limited range of memory types. Naccache, *Quatre exercices de pensée juive pour cerveaux réfléchis,* chaps. 1 and 2.

67. Two varieties of memory accessible to conscious awareness are episodic memory, referring to stored representations of personally experienced events (e.g., remembering that you ate dinner at Two Oaks restaurant in June last year), and semantic memory, referring to stored representations of meaningful facts or general knowledge unrelated to any personal context (e.g., remembering the concept of "dinner" but not a particular dining experience). Episodic memory involves awareness of a sense of having personally experienced an event or item, whereas semantic memory involves awareness of meaning unaccompanied by a sense of familiarity. Lucas, "Memory, Overview."

68. Mirror neurons activated by music are sometimes called echo neurons.

69. The brain's extensive feedback circuitry that monitors the body's actions can only be fully engaged by performing—as opposed to imaging or thinking about—an action.

70. Here the word *mind* does not reflect a medical or legal definition of the word.

71. In Duke Ellington's *Don't Get Around Much Anymore* (1942), the jilted lover who sings "Why stir up memories?" doesn't dread the memories so much as the feelings that accompany them.

72. I realized years later that the lyrics, which speak about unrequited love, were more appropriate for Clapton's 1992 slower-paced and soulful acoustic guitar version, but such considerations were no concern of mine during college parties

73. Salakka et al., "What Makes Music Memorable?," 3.

74. Jacobsen et al., "Why Musical Memory Can Be Preserved in Advanced Alzheimer's Disease," 2438. The anterior cingulate gyrus is also sometimes called the anterior cingulate cortex. I have abbreviated it as ACG for ease of reading. In addition to the ACG, another key area is the ventral pre-supplementary area, a component of the motor system important for encoding rhythm.

75. A full understanding of how the ACG functions remains a work in progress. Devinsky, Morrell, and Vogt, "Contributions of Anterior Cingulate Cortex to Behaviour," 298, conclude that the anterior cingulate cortex and its connections provide mechanisms by which affect and intellect can be joined: "The cingulate gyrus may be viewed as both an amplifier and filter, interconnecting the emotional and cognitive components of the mind." Meanwhile, Vassena, Holroyd, and Alexander, "Computational Models of Anterior Cingulate Cortex," 5–6, identify the ACG as a major brain hub for calculating adaptive behavioral responses to a rapidly changing environment. The ACG has important connections to the reward prediction error system to perform this role.

76. El Haj, Fasotti, and Allain, "The Involuntary Nature of Music-Evoked Autobiographical Memories in Alzheimer's Disease," 239. As an aside, we "retrace over our steps" when we can't remember something via the internal cuing pathway because we are searching for external cues to jog our memory.

77. Salakka et al., "What Makes Music Memorable?," 3. The authors state, "The autobiographical saliency of the music has . . . been reported to be highest in songs popular during the teenage years of the listener." From personal experience, I would also include early adult years.

78. On leveraging MEAM to benefit people with dementia, see chapter 5.

79. Kraus, *Of Sound Mind*, 53.

80. Research has shown that music can even influence visual perception via its effect on mood. Jolij and Meurs, "Music Alters Visual Perception," 4.

3. The Brain's Innate Capacities for Music

1. Allen, *The Lives of the Brain*, 183.

2. This concept, known as redundancy, will be looked at further in the context of recovery after stroke.

3. Typically derived from reports of epileptics during their seizures, this concept is also the one that Wilder Penfield leveraged when he systematically activated sections of brain during surgery to identify the role played by each section. Penfield accomplished this by applying mild electrical stimulation to areas on the surface of the brain.

4. Think of a simple flashlight battery with a positive and a negative end: the movement of ions from one terminal to the other via the light bulb is what generates the electricity that turns on the bulb.

5. One of my favorite souvenirs from my neurology training days is an EEG of my brain waves. It's a physically impressive $13.5 \times 12 \times 1$-inch paper printout weighing nearly 1.5 pounds. That's how EEGs were done in the "old days." Nowadays, a digital EEG takes up essentially no physical space and weighs virtually nothing. The machines used to obtain an EEG can readily be wheeled on a cart to wherever the person being tested is. Digital-era EEGs are tremendously flexible tools, but they don't receive the oohs and aahs that MRI scans do. That's because EEGs don't produce striking visual images of the brain. Yet compared with an MRI scan, an EEG is far less expensive. EEGs are also highly time-sensitive, meaning they excel at measuring the time between a stimulus and a response.

6. Wolff's demonstration is cited in Adachi and Trehub, "Musical Lives of Infants," 230. Watch a baby respond to its dad singing at Reader's Digest, "Newborn Baby Mimics Dad's Voice and Makes Him Giggle."

7. Hepper, "An Examination of Fetal Learning before and after Birth." 95.

8. "Western infants detect a one-semitone change in the context of a Western melody." Adachi and Trehub, "Musical Lives of Infants," 232.

9. For a brief demonstration, see "Consonant vs. Dissonant Intervals: Ear Training,"https://www.youtube.com/watch?v=fabeLu4-Ja0. Whether sounds are pleasing or unpleasant is, of course, a subjective definition. Moreover, dissonance plays a useful role in many types of music, as it generates tension to move a piece forward, while consonance resolves that tension. Thus, it's the proportion of dissonance to consonance and the sensitive placement of dissonant tension that matter most in musical composition.

Consonance and dissonance can be described mathematically: two sounds are consonant if the ratio of their wavelengths is simple whereas two sounds are dissonant if that ratio is complex. Examples of simple, or consonant, ratios are 4:3 (perfect fourth, C→F) and 3:2 (perfect fifth, C→G); the archetypal example of a complex, or dissonant, ratio is 45:32 (the tritone, C→F#). See Trehub, "The Developmental Origins of Musicality," 670. The tritone was considered so dissonant that it was labeled the "Devil's interval" and shunned for centuries in Western music (Paul Nickolloff, personal communication, 2019). The above ratios hold for "just" temperament, the venerable tuning method already familiar in Classical Greece. Modern tuning by "equal" temperament closely approximates these ratios but is not an exact match.

For an example of a young child's distaste for dissonance, see "Baby's Response to Horrible Singing."

10. Masataka, "Preference for Consonance over Dissonance by Hearing Newborns of Deaf Parents and of Hearing Parents," 46.

11. Trehub, "The Developmental Origins of Musicality," 669. Infants likely have this ability thanks to genetically determined cells for relative pitch detection (see chapter 2).

12. Soley and Hannon, "Infants Prefer the Musical Meter of Their Own Culture: A Cross-Cultural Comparison," 287.

13. Trehub, "Infant Musicality," 388. Sandra Trehub (1938–2023) was a pioneer in the study of the musical lives of infants. I had the good fortune to speak with her while researching this material.

14. "The researchers found that the moms' arousal levels were higher during playful compared to soothing song. And they found coordinated decreases in arousal for both the moms and babies as the soothing songs progressed." Cognitive Neuroscience Society, "From Lullabies to Live Concerts: How Music and Rhythm Shape Our Social Brains," March 27, 2018.

15. Bomzer, "Johannes Brahms–Lullaby (Kalima Tab)."

16. Observe the wide range of reactions to music being sung to the baby in "Baby Gets Emotional When Mom Sings Opera." Watch the baby saddened by classical music in "Baby Findley Gets Emotional Hearing Classical Music for the First Time." Recall that Copland wrote that melody is instinctive and he stressed the importance of emotional response or reaction to melody.

17. Trehub, "Musical Universals: Perspectives from Infancy," 7.

18. Watch a four-month-old pay rapt attention to singing in "4 Month Old Baby Tries to Sing to Karen Carpenter Song, Melts Hearts." and three babies with differing reactions, but all paying attention, to the viola at "Babies Listening for the First Time Classical Music Instrument Viola by Boryana Bekirska."

19. Trehub, "Infant Musicality," 389.

20. Zentner and Eerola, "Rhythmic Engagement with Music in Infancy," 5768.

21. See a young child respond to rhythm by dancing in "Baby Reacts to Music."

22. Zatorre, "Music, the Food of Neuroscience?," 314. The observation that babies prefer song over speech is a powerful argument against the claim that music is simply a by-product of language.

23. As previously noted, the maximalist perspective to studying the brain's music functions (highlighted in chapters 1 and 2) focuses on examining normal brains, maximizing the expanse of brain identified as participating in music.

24. *Presbycusis*, the medical term for age-related hearing loss, is a far more common cause of hallucinated sounds than musical partial seizures. Hallucinated sounds from presbycusis are not seizures at all since problems of the brain don't cause them. Rather, they stem from decreased functioning of the ear's hearing structures. I recall a woman, advanced in age, who came to see me with this problem. She had been a professional pianist and related that she would unpredictably hear classical music that she generally liked; her problem was that she couldn't turn her "sound system" off. Testing showed no evidence of seizures or a brain lesion. The problem was due to her declining ability to hear. Some experts conceive of this as a "release phenomenon," meaning that the brain, no longer receiving sufficient hearing input from the external world, creates its own internal soundscape.

Many causes of musical hallucinations have been described. They range from medication side effects and psychoactive drugs to the presence of shrapnel in the cranial vault (as in the intriguing case of Russian composer Dmitri Shostakovich), and from hearing loss to the simply inexplicable. Oliver Sacks devoted the entirety of chapter 6 of *Musicophilia* to this fascinating topic.

25. MacDonald Critchley, coauthor of an acclaimed book about the neurology of music, describes the first musicogenic seizure he witnessed as follows: "Not many bars [of Tchaikowsky's "Waltz of the Flowers"] elapsed before the patient began to look distressed, and gradually she developed a seizure with generalized convulsive movements, frothing at the lips, and cyanosis . . . The patient subsequently told me, 'That's the sort of music which always brings on an attack.'" Critchley and Henson, *Music and the Brain: Studies in the Neurology of Music*, 345. In medical terms, this patient experienced a focal seizure with secondary generalization.

26. Brust, "Music and the Neurologist," 144. While agreeing with the temporal lobe location, a recent review of musicogenic seizures stressed the wide diversity of underlying causes of these seizures and presented case studies of two patients who had no musical training. Bratu et al., "Musicogenic Seizures in Temporal Lobe Epilepsy: Case Reports Based on Ictal Source Localization Analysis," 2.

27. Brust, "Music and the Neurologist," 144.

28. This case illustrates the common prioritizing by medical professionals of language over music: the locations of language skills are routinely assessed prior to such surgery whereas music skills are not. The takeaway lesson is, as the authors of the published patient report conclude, "Patients about to undergo a frontal lobe resection [operation] should be queried regarding [asked about] the importance of musical expression in their lives."

The lesson from the above report appears to have been learned. While researching this book, I came across a cool post on LinkedIn from June 2024 that described a musician who was woken up to play the violin midway through brain surgery in order to ensure that the parts of the brain responsible for intricate hand movements were not affected by the operation. Per the post, the patient's tumor was located in the right frontal lobe.

29. Per the case reported in McChesney-Atkins et al., "Amusia after Right Frontal Resection for Epilepsy with Singing Seizures," testing after surgery showed that "Ben" had no **aprosody**, meaning that his ability to perceive and express the sentimental contents of speech remained normal. This indicates that prosody occurs in areas of the brain distinct from the (expressive musical skills) part of Ben's brain removed at surgery. Prosody is a complex process and can be based on multiple inputs, including visual. Garrido-Vásquez et al., "Dynamic Facial Expressions Prime the Processing of Emotional Prosody," 1.

30. A study of early post-stroke amusia patients acknowledged as much, noting, "Because of rapid post-lesional reorganization . . . musical functions were assessed between the third and seventh day [post-stroke]." Rosslau et al., "Clinical Investigations of Receptive and Expressive Musical Functions after Stroke," 4.

31. A classic example of this phenomenon is called left-sided neglect. Many patients who suffer a stroke in the right hemisphere will pay no attention to the left side of their body or of the environment when they present to the hospital. It's quite remarkable to observe a patient for whom the left side of the world doesn't exist. Amazingly, left-sided neglect usually clears up on its own within three months and often sooner than that. Only the largest strokes of the right hemisphere cause permanent left-sided neglect.

32. Sihvonen et al., "Neural Architectures of Music—Insights from Acquired Amusia," 111.

33. Sihvonen et al., "Neural Basis of Acquired Amusia and Its Recovery after Stroke," 8875: "41% of amusics had concurrent aphasia, 65% of aphasics had concurrent amusia."

34. Beat deafness, a congenital disturbance of rhythm perception or expression, has also been described. See Dalla Bella and Sowinski, "Uncovering Beat Deafness," 1–11.

35. Tone deafness may be receptive (a decreased ability to perceive the difference between two closely related notes) or expressive (a decreased ability to produce a pitch change between two closely related notes). The

example presented in association with Figure 3.2 describes receptive tone deafness.

36. The chromatic scale is A, A#/B♭, B, C, C#/D♭, D, D#/E♭, E, F, F#/G♭, G, and G#/A♭. A slash (as in A#/B♭, for example) means this is the same note and may be called either A-sharp or B-flat.

37. Adachi and Trehub, "Musical Lives of Infants,' 232. Humans are born with cells specialized for relative pitch discrimination (see chapter 2).

38. Loui, Wan, and Schlaug, "Neurological Bases of Musical Disorders and Their Implications for Stroke Recovery," 30. Peretz, *How Music Sculpts the Brain*, 50, notes that in tone deafness, the gray matter to which the arcuate fasciculus connects at each end is also abnormal. This raises the question of which abnormality—the white matter or the gray matter—is cause and which is effect. Peretz relates the condition to an underlying genetic abnormality.

39. Sihvonen's research group reported the music network in appendix A. His research interests also include the rehabilitative use of music for neurological patients. For more on his work, see Aleksi Sihvonen, ResearchGate, https://www.researchgate.net/profile/Aleksi-Sihvonen.

40. A limitation of meta-analysis is that different studies are performed differently, so questions arise about how to combine the data from multiple studies. It's not to the extent of comparing apples and oranges; rather, it's more like comparing Delicious apples with McIntosh apples.

41. Griffiths is a leader in the field of the cognitive aspects of human hearing.

To view more about him, see the Academy of Medical Sciences, https://acmedsci.ac.uk/fellows/fellows-directory/ordinary-fellows/fellow/Professor-Timothy-Griffiths-0008981.

42. Shebalin's formal diagnosis was a Wernicke's (receptive i.e., difficulty comprehending speech) aphasia. Wernicke's area and Broca's area are located in the left hemisphere of the brain (see figure 2.1).

43. Segelman, "Vissarion Shebalin." https://www.wisemusicclassical.com/composer/2984/Vissarion-Shebalin/.

44. Leborgne represents the index case of what is now called Broca's (expressive, i.e., difficulty producing speech) aphasia.

45. Brown directs the NeuroArts Lab. To view more about him, see "Steven Brown," Research and Innovation: Experts, McMaster University, https://experts.mcmaster.ca/display/stebro.

46. On-demand blood flow serves as a mechanism for the brain to conserve energy.

47. This is why the brain template (see figure 3.3) is drawn with a proportionately larger temporal lobe.

48. Input to the auditory cortex of both brain hemispheres is bilateral, meaning that each hemisphere's auditory cortex receives sound input from both ears.

49. Brown, Martinez, and Parsons, "Music and Language Side by Side in the Brain: A PET Study of the Generation of Melodies and Sentences," 2800. Couched in technical language, the precise quote reads, "The distinct non-overlapping, domain-specific activations . . . observed for the melody and sentence generation tasks may be due to operationalized task-related differences in their informational content (semantics)."

50. These quotes are found in Eichler, *Time's Echo: The Second World War, the Holocaust, and the Music of Remembrance*, 246

51. McQuay, "How Sound Shaped the Evolution of Your Brain."

52. I use the terms *frequency discrimination* and *spectral discrimination* interchangeably.

53. Werner Heisenberg developed the uncertainty principle to describe subatomic particles, such as photons or electrons.

This classic joke, which I paraphrased from Homick, *From Time to Time*, 39, humorously illustrates Heisenberg's principle that it's not possible to have full knowledge of location (position) and velocity simultaneously: Heisenberg is driving a car and gets pulled over. The cop asks him, "Do you know how fast you were going?" "No," Heisenberg replies, "but I know exactly where I am." The cop says, "You were doing 55 on a 35 miles per hour road." Heisenberg throws up his hands and exclaims, "Great! Now I'm lost!"

54. Zatorre, Belin, and Penhune, "Structure and Function of Auditory Cortex: Music and Speech," 41: "The 'Acoustic Uncertainty Principle,' explicitly articulated by [Martin] Joss, states that one cannot make a precise simultaneous measurement of an auditory event in both the time and frequency domains."

55. Washington and Tillinghast, "Conjugating Time and Frequency: Hemispheric Specialization, Acoustic Uncertainty, and the Mustached Bat," 1. An advanced mathematician told me that complementary variables in the case of sound—frequency and time—are technically called canonically conjugate variables, variables that are Fourier transformations of one another.

Zatorre, *From Perception to Pleasure: The Neuroscience of Music and Why We Love It*, 155, notes, "Complementary hemispheric specialization for spectral and temporal processing . . . does not depend on the stimuli being semantically meaningful." Rather, semantic meaning for music and language was made possible through leveraging the complementary hemispheric specialization that had developed in response to the physics of sound consistent with the Joos uncertainty principle. This accords with the Darwinian principles I recall from high school biology. Per Darwin, animals which happened to have longer necks could reach higher fruit, and this advantage was passed on to future generations, leading to modern giraffes. Per Lamarck, seeing the fruit high up in the trees, animals stretched and stretched their necks, with modern giraffes resulting over generations of neck stretching.

56. Zatorre, "Neuronal Specializations for Tonal Processing," 239. For a more detailed diagram, see Zatorre, *From Perception to Pleasure*, 164–67, fig. 5.8. I wish to acknowledge a personal communication from Robert Zatorre for this reference.

57. Zatorre, Belin, and Penhune, "Structure and Function of Auditory Cortex: Music and Speech," 37.

4. Music's Evolutionary Benefit to Humans

1. Fitch, "The Evolution of Music in Comparative Perspective," 3. Couched in scientific terminology, Fitch states that music is appositional whereas language is propositional.

2. Levitin, *The World in Six Songs*, 137–87.

3. Levitin, 159.

4. Levitin, 162.

5. Tomaino, *Music Has Power® in Senior Wellness and Healthcare*, 46.

6. Levitin, *The World in Six Songs*, 168.

7. Levitin, 159.

8. Levitin, 64. Although the English word *ballad* derives from the Latin verb *ballare*, which means "to dance," I don't suspect that Levitin associates the term with dance given how he uses it elsewhere in his book, such as labeling Bob Dylan's song "Hurricane" a ballad.

9. Nash, "Folk-Pop Legend Gordon Lightfoot Goes Solo."

10. I use the Canadian term *First Nations* because Lightfoot was Canadian.

11. Its actual name is the Mariners' Church of Detroit. Over the years, its bell tolled twenty-nine times each November 10 in memory of the crew members lost.Beginning on November 10, 2023, its bell began tolling thirty times to acknowledge Gordon Lightfoot after his death on May 1, 2023.

12. "La Bohème," music by Aznavour and lyrics by Jacques Plante, resonates deeply for me. In the preface, I recalled my high school French teacher. Learning from her set in motion a series of events in my life that

led to my coming to know this song, which drew me into the mystique of Montmartre and stimulated me to learn about its fabled past.

Aznavour was the greatest male singer of the French chanson music era, which stretched more or less from the last decade of the nineteenth century through the 1960s and was roughly contemporaneous with the American popular songbook era. Aznavour sang about many different themes: finding love, oh for sure; losing love, oh the heartbreak; Paris, oh là là. He also sang many songs about youth, which is to say about personal growth through the trials and tribulations of the young adult years. "La Bohème" expresses all of these themes. Aznavour considered himself, first and foremost, a practitioner of the art of telling stories through song. See Riding, "At 82, Charles Aznavour Is Singing a Farewell That Could Last for Years."

13. Before my wife and I became parents, we rented an apartment on the edge of Montmartre for a short time. From various walking tours, we learned about the history and feel of the neighborhood.

14. Gould, *The Fate of Carmen*, 28–31.

15. Macmillan, "A Self-Guided Walk on the Artists' Trail in Montmartre," https://montmartrefootsteps.com/montmartre-historical-cultural-context/.

16. Trehub, Becker, and Morley, "Cross-Cultural Perspectives on Music and Musicality," 4.

17. Grieser and Kuhl, "Maternal Speech to Infants in a Tonal Language: Support for Universal Prosodic Features in Motherese," 14.

18. Masataka, "Motherese in a Signed Language," 453.

19. Saint-Georges et al., "Motherese in Interaction," 3.

20. Saint-Georges et al., 9–10.

21. Saint-Georges et al., 11.

22. I learned this song from a senior physician during a pediatric neurology rotation when I was in training. He used it to calm crying babies so he could examine them.

23. Fitch, "The Evolution of Music in Comparative Perspective," 7. Some species of gibbons can sing, but they have no capacity for vocal learning. Gibbons can vocalize a sequence of notes that respect a time constraint so are said to sing. The male and the female of a monogamous pair (gibbons are one of the few mammalian species that are monogamous) sing duets in time with one another. See Dinneen, "Male and female gibbons sing duets in time with each other," https://www.newscientist.com/article/2354278 -male-and-female-gibbons-sing-duets-in-time-with-each-other/. However, gibbons do not possess the capacity of vocal learning so these are genet-

ically determined calls, rather than songs. Geissmann, "Gibbon Songs and Human Music from an Evolutionary Perspective," 105. For an informative discussion of whale music, see Spitzer, *The Musical Human*, 297–304.

24. Mehr et al., "Form and Function in Human Song," 356. Dance songs came in second at 90 percent, healing songs third, at 71 percent. Love songs were the least distinctive type, recognized at about the same level as chance. This study included an additional, fascinating survey: More than nine hundred academics from three disciplines were invited to predict the results of the study. This survey revealed that 73 percent of cognitive scientists predicted that listeners (the study participants) would make accurate inferences concerning the assorted song types, whereas 52 percent of music therapists and only 29 percent of ethnomusicologists predicted that listeners would make accurate inferences. The gulf between cognitive scientists at 73 percent and ethnomusicologists at 29 percent is truly striking. The study authors did not offer a possible explanation for this difference. Might it reflect the orientations, or biases, of their professions, specifically, that cognitive scientists observe how the brain processes music similarly—its innate capacities for music—regardless of a particular brain's upbringing whereas ethnomusicologists have as their point of departure the variations of musical forms among different cultures—the learned aspects of music? This reminds me of the different mindsets maximalists and minimalists bring to studying scientific data.

25. Trehub, "Musical Universals: Perspectives from Infancy," 7.

26. For an interesting exchange on the song's origin, see Mudcat Café, "Origin: If You're Happy and You Know It- How old?" https://mudcat.org/thread.cfm?threadid=20648.

27. Thaut, "The Discovery of Human Auditory–Motor Entrainment and Its Role in the Development of Neurologic Music Therapy," 254.

28. Speech, being neither metrically regular nor constrained by time, does not result in entrainment.

29. Zentner and Eerola, "Rhythmic Engagement with Music in Infancy," 5768.

30. Trehub, "Infant Musicality," 390.

31. Trehub, Becker, and Morley, "Cross-cultural Perspectives on Music and Musicality," 4–5.

32. Winsler, "Singing One's Way to Self-Regulation," 290. Kindermusik™ specializes in research-based music and movement curricula and classes.

33. Launay, "Choir Singing Improves Health, Happiness—and Is the Perfect Icebreaker."

34. The Vault, "10 Lessons Your Child Will Learn in Marching Band."

35. Trehub, Becker, and Morley, "Cross-cultural Perspectives on Music and Musicality," 4.

36. I say this from personal experience volunteering for this task. View the video "Donkey Riding - *with capstan demonstration*."

37. Harlow, *Chanteying Aboard American Ships*, 1.

38. Harlow, 92–93. Harlow's book is also an ethnography of sorts as he occasionally describes differences in the music and lyrics used by American and Afro-Caribbean sailors.

39. See chapter 1.

40. Zentner and Eerola, "Rhythmic Engagement with Music in Infancy," 5771. The authors acknowledge that it takes a degree of motor control maturation for children to be able to match the timing of the rhythm they hear precisely with the timing of their body movements.

41. Trainor et al., "The Primal Role of the Vestibular System in Determining Musical Rhythm," 41.

42. Trainor et al., "The Primal Role of the Vestibular System in Determining Musical Rhythm," 37.

43. Kraus, *Of Sound Mind: How Our Brain Constructs a Meaningful Sonic World*, 51. Kraus further remarks, "The ear [i.e., hearing] arose from organs designed to perceive gravity and an organism's place in space with the goal of achieving movement."

44. Trainor directs the Institute for Music & the Mind at McMaster University in Hamilton, Ontario, Canada. To view more about her, visit McMaster University, Experts, https://experts.mcmaster.ca/display/ljt.

45. Trainor et al., "The Primal Role of the Vestibular System in Determining Musical Rhythm," 37 and 41.

46. Todd and Lee, "The Sensory-Motor Theory of Rhythm and Beat Induction 20 Years On: A New Synthesis and Future Perspectives," 12–18.

47. "African American Spirituals," Library of Congress, https://www.loc.gov/item/ihas.200197495/.

48. Guided tours are available by reservation. For information about the park, visit Harriet Tubman Home, https://www.harriettubmanhome.com/.

49. For details on the sculpture and its sculptor, visit Wesley Wofford–FNSS/Wofford Sculpture Studio, https://www.woffordsculpturestudio.com/exhibitions/harriet-tubman.

50. Song of Songs 2:10–12 (*Tanakh: The Holy Scriptures* (New Translation), 1407). In one of the better-known examples of biblical translation errors, multiple English versions mistakenly translated the Hebrew word *toor*

as "turtle" rather than "turtledove." Members of a band that specialized in Sephardic (Judeo-Spanish) music embraced this mistranslation and named themselves Voice of the Turtle.

51. Song of Songs 5:6 (*The Holy Bible*, Revised Standard Version), 597.

52. Jordan, "Four Ancient Chinese Songs."

53. Black, "Ancient Egyptian Love Songs Connect Romance Across Time."

54. From Papyrus Harris 500, 18th Dynasty, British Museum, London, acc. no. EA10060; Guzzo, "Hiddenness and Darkness in Ancient Egyptian Love Songs."

55. According to Kraus, *Of Sound Mind*, 190: "Like much human singing, [birdsong] is all, or at least mostly, about sex."

56. The phrase "language of birds" celebrates the achievements of the great poets and writers—Dante, Milton, Shakespeare, Cervantes, Rabelais, and so on—referring to their ability to "sing" with language as beautifully and naturally as birds do. (Allan Graubard, personal communication, 2023)

57. Klein, "Some Songbirds Have Brains Specially Designed to Find Mates for Life."

58. Moore and Woolley, "Emergent Tuning for Learned Vocalizations in Auditory Cortex," 1470.

59. Geschwind, "Neurological Knowledge and Complex Behaviors," 191–93. Geschwind's concept is consistent with the dynamic tension between localization and equipotentiality presented in chapter 2.

60. Gibbons can vocalize a sequence of notes that respect a time constraint so are said to sing. The male and the female of a monogamous pair sing duets in time with one another; they are one of the few mammalian species that are monogamous. See Dinneen, "Male and Female Gibbons Sing Duets in Time with Each Other." Gibbons do not possess the capacity for vocal learning so their sounds are genetically determined calls, rather than songs. Geissmann, "Gibbon Songs and Human Music from an Evolutionary Perspective," 105.

61. Music itself is likely an example of convergent evolution. This would explain why bird music shares some but not all attributes of human music, and why gibbon music shares different attributes of human music but not others. Since birds are found throughout the world, whereas gibbons are limited to the forests of Southeast Asia, bird singing has had a much greater impact on human concepts about music. Lots of song lyrics evoke birds, but do any describe gibbons? Actually, here's one: "Sad the calls of the gibbons at the three gorges of Pa-tung/After three calls in the night, tears wet the

traveler's dress." This fourth-century Chinese song shows that gibbon songs evoke human emotional responses. Geissmann, "Gibbon Songs and Human Music from an Evolutionary Perspective," 103.

62. Cepelewicz, "In Birds' Songs, Brains and Genes, He Finds Clues to Speech." Of note, "Songbirds make up almost half of the existing 9,000 avian species." Whaling, "What's Behind a Song?," 65.

63. The argument for neurogenesis—that new neurons can be added to the brains of adult higher vertebrates like birds and mammals—dates from the 1960s. Nonetheless, the conventional wisdom that no new brain cells form after birth in mammals persisted for several decades. The concept of neurogenesis in adults did not secure general recognition until approximately forty years later.

Neurogenesis is one of multiple points of convergence between human and bird singing described by Perrone-Capano, Volpicelli, and di Porzio, "Biological Bases of Human Musicality," 2. Another is expression of the FOXP2 gene by both songbirds and humans. Mutations of this gene in humans cause severe pronunciation impairments in childhood, and ablation of this gene in birds prevents song learning.

64. Brenowitz and Larson, "Neurogenesis in the Adult Avian Song-control System," 3. The authors point out that brain regions capable of neurogenesis are more limited in humans compared to songbirds—just the hippocampus and olfactory bulb in humans; throughout most of the forebrain in songbirds.

65. Harding, "Learning from Bird Brains: How the Study of Songbird Brains Revolutionized Neuroscience," 32.

66. Brenowitz and Larson, "Neurogenesis in the Adult Avian Song-Control System," 1.

67. Helmuth, "Why Bird Brains Bloom in Spring."

68. Loui, Wan, and Schlaug, "Neurological Bases of Musical Disorders and Their Implications for Stroke Recovery," 28.

69. This was strikingly evident in the early days of rock 'n' roll. Just take a look at videos of performances by Elvis Presley or the Beatles.

70. "Different Drum," recorded by Ronstadt as a member of the Stone Poneys, written in 1964 by Michael Nesmith (of Monkees fame); "It's So Easy to Fall in Love," a Buddy Holly tune from 1956; "Hurt So Bad," written by Teddy Randazzo, Bobby Weinstein, and Bobby Hart in 1964, originally a hit for Little Anthony & the Imperials; "When Will I Be Loved?," originally a minor hit for the Everly Brothers (songwriters Phil and Don) in 1960.

71. Written by band member Christine McVie in 1975.

72. Museum of Fine Arts, Boston, https://collections.mfa.org/objects /32592.

73. Contrast this pose to formal closed position in ballroom dancing. See BallroomDancers.com, https://www.ballroomdancers.com/Learning _Center/Technique/Dance_Positions/info.asp?pos=cp.

74. Smee, "MFA Expands Loans of Well-Known Works."

75. No written description can do this courtship display justice. It needs to be seen to be appreciated. Luckily, it can be viewed at Idaho Fish and Game, "Dance of the Sage Grouse," https://www.youtube.com/watch?v =28D491RoBEs. Watch it and then decide for yourself what role dance plays in sexual selection for sage grouses.

76. Todd and Lee, "The Sensory-Motor Theory of Rhythm and Beat Induction 20 Years On: A New Synthesis and Future Perspectives," 1.

77. The limbic brain's communication with the vestibular system by way of the pathway for external guidance allows for current movements to be rewarded.

78. Todd and Lee, 17.

79. Copland, *What to Listen for in Music*, 34.

80. Chatterjee, *The Aesthetic Brain*, 34–35. Chatterjee details that in concert with the vestibular system, the basal ganglia and the cerebellum are also deeply involved with coordinating the body's motor response action plans. These regions, along with cortical areas responsible for processing body movement, are richly involved with dance.

81. Take a look at videos of Elvis Presley performing or view any of the lavish song-and-dance numbers in Bollywood wedding scenes.

5. Improved Quality of Life through Music

1. On music's evolutionary benefit to the brain, see chapter 4. On the brain's innate capacities for music, see chapter 3.

2. The metrics of survival and quality of life correlate with natural selection. Studies and articles (and experience) reveal that music enhances sexual arousal and performance. A clinically meaningful impact on sexual selection, however, would further require demonstrating increased procreation success. Studying this methodically in vivo would present considerable challenges, to say the least. On the other hand, the Institut Marquès in Spain has studied the issue in vitro and reports that the presence of music in its embryo incubators increases the fertilization rate by 5 percent. The researchers attribute this effect to musical vibrations producing "movements similar to those that fertilised eggs experience in their journey

through the fallopian tubes and the uterus." For more details, visit Institut Marquès, "Music Improves In-Vitro Fertilization," https://institutomarques.com/.

3. Altenmüller and Schlaug, "Apollo's Gift," 240.

4. Schlaug, "Musicians and Music Making as a Model for the Study of Brain Plasticity," 40.

5. Yun, "Music, Rhythm, and the Brain."

6. Groussard et al., "The Effects of Musical Practice on Structural Plasticity," 174.

7. Schlaug, "Musicians and Music Making as a Model for the Study of Brain Plasticity," 39.

8. Balbag, Pedersen, and Gatz, "Playing a Musical Instrument as a Protective Factor Against Dementia and Cognitive Impairment," 3. The study revealed similar findings whether the twins were identical or fraternal, supporting the view that this is a learning effect (nurture) rather than a genetic effect (nature).

9. Curry, "Imaging the Past," 43–44. In a second part of the study, the volunteers were divided into two groups, with half given verbal and visual instruction and the other half given only visual instruction. The volunteers' brains were then scanned as they made the tools. Results showed a difference between the two groups. Language brain regions activated in those who had received verbal instruction. For those who received visual instruction only, brain regions used to make music activated while language regions remained quiet. Scientists at the Stone Age Institute, in Bloomington, Indiana, concluded that complex toolmaking could have appeared independently of the appearance of language by virtue of the presence of multiple brain pathways that can be recruited to accomplish and pass on the knowledge of this function.

10. Rogenmoser et al., "Keeping Brains Young with Making Music," 297.

11. Hanna-Pladdy and MacKay, "The Relation Between Instrumental Music Activity and Cognitive Aging," 378.

12. Hanna-Pladdy and MacKay, 384. The concept of cognitive reserve is generally credited to Katzman, "Education and the Prevalence of Dementia and Alzheimer's Disease," 18, although he used the term *brain reserve* in the article. Katzman was visionary in anticipating the profound impact that Alzheimer's disease would come to have on global health care.

13. Music in the context of healthy aging is often studied as a component of "leisure activities." The classic article on leisure activities and pres-

ervation of brain function is Verghese et al., "Leisure Activities and the Risk of Dementia in the Elderly," 2003.

14. Klimova, Valis, and Kuca, "Cognitive Decline in Normal Aging and Its Prevention: A Review on Non-Pharmacological Lifestyle Strategies," 905.

15. Chapter 6 explores a variety of ways to add music to one's life for anyone not already actively engaged with it.

16. Sihvonen et al., "Music-Based Interventions in Neurological Rehabilitation," 649.

17. Altenmüller and Schlaug, "Apollo's Gift," 245.

18. A patient's motivation impacts the ultimate success of any rehabilitation approach.

19. Grau-Sánchez et al., "Music-Supported Therapy in the Rehabilitation of Subacute Stroke Patients," e191.

20. Observe in figure 1.3 how closely the somatosensory cortex, which processes proprioception, is located to the motor cortex. As a medical aside, motor pathways serve components beyond strength, muscle tone being a prime example.

21. Schmitz et al., "Movement Sonification in Stroke Rehabilitation," 2–3.

22. Scholz et al., "Sonification of Arm Movements in Stroke Rehabilitation," 1.

23. Techniques using visually augmented reality, sometimes as simple as the strategic placement of a mirror, are also being developed and utilized.

24. Although the cause or causes of the encephalitis lethargica epidemic have not been unequivocally elucidated, that its onset coincided with the Spanish flu pandemic which began in 1918, strongly suggests the flu's link to at least some of the cases. I have a personal connection to this since both of my father's parents were widowed by the Spanish flu. They subsequently married one another. So, I can honestly say that were it not for the Spanish flu pandemic, I would not be here.

25. Carl's smile in the patient vignette earlier in this chapter is an example of this. When he smiles volitionally, his smile is asymmetric, weaker on the paralyzed side because of his stroke, but when he smiles by emotional reflex, his smile is full and symmetric, revealing that the neural pathways for this action bypass the brain region affected by his stroke.

26. Sacks, *Musicophilia*, 249.

27. Chen, Penhune, and Zatorre, "Listening to Musical Rhythms Recruits Motor Regions of the Brain," 2844. The first part of the study showed activation of multiple motor areas of the brain—supplementary motor area,

mid-premotor cortex, and cerebellum—by simply listening to musical rhythms. The experimenters then separated the participants into two groups. The first group was told they would have to tap the rhythm after listening to it, but the second group was not told this. The same motor regions were recruited in both groups with one key difference: The first group also activated the anterior premotor cortex in association with their thinking about tapping the rhythm.

28. Ashoori, Eagleman, and Jankovic, "Effects of Auditory Rhythm and Music on Gait Disturbances in Parkinson's Disease," 7. Videos about rhythmic auditory stimulation can be found online. See, for example, Gunlock, "Gait Training for Parkinsons's [sic] Patient Using Music," https://www .youtube.com/watch?v=uDjQ7lKmH3s.

29. Mainka et. al., "The Use of Rhythmic Auditory Stimulation to Optimize Treadmill Training for Stroke Patients," 2.

30. Sihvonen et al., "Music-Based Interventions in Neurological Rehabilitation." 649.

31. Davis, Gfeller, and Thaut, *An Introduction to Music Therapy Theory and Practice*, 286.

32. Davis, Gfeller, and Thaut, 287.

33. It's common knowledge that music can take our problems "off of our mind" while engaging with it. This vignette illustrates a concept of pain known as gate theory. Simply stated, the brain can receive and process only a limited bandwidth of sensory information from the body at any given time. The mechanism controlling what signals reach or do not reach the brain is called a "gate." Gates exist throughout the nervous system, in peripheral nerves, the spinal cord, and the subconscious brain. The more pain signals that pass the gates and reach the brain, the more they will come into awareness and a person experiences pain. If other types of signals can pass through the gates, however, there will be less bandwidth available for pain signals to pass. Engaging this patient with music favored body sensations related to melody and rhythm. The music therapist was thus able to reduce the bandwidth of pain signals passing the gates and reaching awareness. This reduced the patient's perception of pain ultimately allowing him greater capacity to move.

The word *brain* in this pain management section refers to the conscious brain—that is, the brain that is aware of pain. Zhou et al., "Sound Induces Analgesia Through Corticothalamic Circuits," 6, report that the thalamus—a major sensory relay station in the subconscious brain—functions as a gate, and its activity can be modulated by sound, such as the sound of music. In this vignette, the music therapist also makes use of other aspects of music beyond sound, such as movement by singing and drumming, modulating additional gates in the nervous system.

34. Konnikova, "The Man Who Couldn't Speak and How He Revolutionized Psychology."

35. On localization, see chapter 2, and on the location of Broca's area, see figure 2.1.

36. Sacks, foreword to *The Paradoxical Brain*, xiv.

37. Johnson and Graziano, "Some Early Cases of Aphasia and the Capacity to Sing," 78–82.

38. For an example of melodic intonation therapy, see MedRhythms, "Melodic Intonation Therapy—Longer Phrases," https://www.youtube.com/watch?v=t0Vv-JHhIsl.

39. Sihvonen et al., "Neural Architectures of Music," 109.

40. He arrived in the emergency department beyond the time window for "clot-busting" (thrombolytic) medication.

41. That this gentleman didn't start MIT until several months after his stroke may also have contributed to his lack of response.

42. For example, see Van der Meulen, Van de Sandt-Koenderman, and Ribbers, "Melodic Intonation Therapy," S50–S51.

43. This serves as a reminder of the bias favoring language assessment over music assessment, similar to what the patient "Ben"—the chorale director who lost musical skills following surgery—encountered in chapter 1.

44. Watson, "Critical Review: Melodic Intonation Therapy," 3.

45. Salakka et al., "What Makes Music Memorable?," 2.

46. Molnar-Szakacs and Overy, "Music and Mirror Neurons," 238.

47. Menke et al., "Family-Centered Music Therapy," 1. The study enrolled sixty-five parent-infant pairs. While the result proved promising, further study with more subjects is needed to demonstrate statistical significance.

48. According to Menke et al., "Family-Centered Music Therapy," 12, "Improvement in the medical condition of the infants has the strongest effect on reducing stress and anxiety of parents."

49. Haslbeck and Bassler, "Music from the Very Beginning," 1.

50. Disease refers to the underlying pathology occurring in the brain. Alzheimer's dementia is thus the clinical condition caused by Alzheimer's disease.

51. For more facts and figures on dementia, visit the Alzheimer's Association, www.alz.org.

52. Jacobsen et al., "Why Musical Memory Can Be Preserved in Advanced Alzheimer's Disease." 2438. The technical terms for the two brain

areas cited in this study are anterior cingulate gyrus (ACG) and ventral presupplementary motor area (VPSMA), respectively.

53. Although it may appear like a single structure, the front (anterior) part and the back (posterior) part of the cingulate gyrus belong to different networks of the brain. The posterior cingulate gyrus is strongly involved in internally cued autobiographic memory and belongs to a network called the default mode. See Maddock, Garrett, and Buonocore, "Remembering Familiar People: The Posterior Cingulate Cortex and Autobiographical Memory Retrieval," 667. The anterior cingulate gyrus (ACG) plays an important role in semantic memory and is believed to play a role in linking the memory to its emotional setting given that the ACG is a component of the limbic system's network. See Kaneda and Osaka, "Role of Anterior Cingulate Cortex During Semantic Coding in Verbal Working Memory," 60. Because the default mode network is a major target of Alzheimer's disease pathology, posterior cingulate gyrus function falters in Alzheimer's dementia. The limbic system, on the other hand, continues to function properly because it is more resistant to AD pathology. See Seeley et al., "Neurodegenerative Diseases Target Large-Scale Human Brain Networks," 46.

54. See Platel et al., "Semantic and Episodic Memory of Music Are Subserved by Distinct Neural Networks," 244. Also, the anterior cingulate gyrus's rich connection within the limbic system imbues semantic memory with emotion.

55. Your intact episodic memory enables you consistently to connect a MEAM to your personal history with the song.

56. The ventral pre-supplementary motor area is associated with preserved muscle memory among individuals with Alzheimer's dementia. Like the ACG, the VPSMA tends to resist the pathology of Alzheimer's disease. The VPSMA plays a role in "acquiring [learning] a novel sequence of movements" and so is linked to procedural memory. See Ohbayashi, "The Roles of the Cortical Motor Areas in Sequential Movements," 4. This finding explains how musically trained individuals with Alzheimer's dementia can learn new pieces of music even if they retain no episodic memory of practicing them. In fact, non-musicians with Alzheimer's can also learn songs, thanks to preserved procedural memory. See Baird, Umbach, and Thompson, "A Nonmusician with Severe Alzheimer's Dementia Learns a New Song," 36.

57. Medically speaking, behavioral disturbances are often called neuropsychiatric symptoms.

58. Sihvonen et al., "Music-Based Interventions in Neurological Rehabilitation," 654. The authors of the paper propose several mechanisms for

music's beneficial effects on neuropsychiatric symptoms: activation of the dopaminergic mesolimbic system (including the reward prediction error system), activation of the parasympathetic system and inhibition of the sympathetic nervous system, and reduced cortisol release with reduced cardiovascular stress (656–57).

59. Platel and Groussard, "Benefits and Limits of Musical Interventions in Pathological Aging," 325.

60. A similar effect is observed with young children. See "Caregiver Hypothesis" in chapter 4.

61. One report showed that dance yields a positive effect on physical and cognitive function as well as on quality of life for individuals with Alzheimer's disease: Ruiz-Muelle and López-Rodríguez, "Dance for People with Alzheimer's Disease," 919. The Cochrane Database attempted a review of dance movement therapy for dementia but did not find any studies meeting their inclusion criteria. Karkou and Meekums, "Dance Movement Therapy for Dementia," 5.

62. This brings to mind young Dr. Sayer's insight in *Awakenings* about distinct volitional (at-will) and non-volitional (reflex) pathways to initiate movement.

63. Fang, Ye, Huangfu, and Calimag, "Music Therapy Is a Potential Intervention for Cognition of Alzheimer's Disease," 6. Särkämö, Tervaniemi, and Huotilainen, "Music Perception and Cognition," 447, reported, "AD individuals had better recognition accuracy for the sung lyrics than the spoken lyrics, whereas healthy controls showed no significant differences." Similarly, Platel and Groussard, "Benefits and Limits of Musical Interventions in Pathological Aging," 326, state that music can be "used as a mnemonic proxy to decrease the difficulties of verbal learning."

64. Särkämö et al., "Cognitive, Emotional, and Social Benefits of Regular Musical Activities in Early Dementia," 643. This effect was particularly noted for remote (long ago) personal memories.

65. Moreira, Reis Justi, and Moreira, "Can Musical Intervention Improve Memory in Alzheimer's Patients? Evidence from a Systematic Review," 137.

66. Numerous videos on MEAM can be accessed online. For example, see Central Michigan University, "Music Therapy Helps Elderly with Alzheimer's and Dementia Recall Long-Term Memories," https://www.youtube.com/watch?v=aKQsxKhtDos.

67. The stimulating effect is inconsistent due to the impairment of the episodic memory of people with Alzheimer's dementia.

68. Fang et al., "Music Therapy Is a Potential Intervention for Cognition of Alzheimer's Disease," 2.

69. Matziorinis and Koelsch, "The Promise of Music Therapy for Alzheimer's Disease," 12. Davis, Gfeller, and Thaut, "Music Therapy: An Introduction to Theory and Practice," 200–201, identify nineteen positive outcomes for music therapy in elderly populations.

70. How each person is a better member is unique; by way of examples, for some people it might be that they respond better in a group, but for others sitting quietly for longer periods of time may constitute their being better members of the group

71. A piece that musically traces the arc of Alzheimer's disease in three movements for 14 instruments, two vocalists, and chorus was composed by Robert Cohen in 2009 with libretto by Herschel Garfein. The first movement ends by focusing on Auguste Dieter's famous quote. For more information, see https://www.wisemusicclassical.com/work/64118/Alzheimers-Stories--Robert-S-Cohen/.

72. Davis, Gfeller, and Thaut, "Music Therapy," 197.

73. Internally cued and externally cued access to one's autobiographic memory operate via different pathways in the brain. So, in the case of MEAM, what mechanism stimulates access, albeit inconsistently, to episodic memory (the kind of memory that is compromised in Alzheimer's dementia)? Could it be that the semantic memory activity occurring in the preserved anterior cingulate gyrus stimulates the functionally impaired posterior cingulate gyrus, thus igniting the autobiographic recall? If so, this further reinforces the importance of preserved anterior cingulate gyrus function for musical interventions in demented individuals.

74. I've also witnessed externally cued autobiographic recall prompted by demented individuals viewing old photographs of themselves and/or people they were close to when they were younger. So, bring along some old pictures too.

6. Coda: Adding Music to Your Life

1. In 1877.

2. Of course, music had for centuries been "for sale" in the form of concerts, with the wealthy paying musicians to perform in their mansions and chateaux. For the vast majority of people, however, having music in their home meant being involved in making it.

3. Music was a component of the Quadrivium as explained by Dr. Shuler in the afterword.

4. Peretz, *How Music Sculpts the Brain*, 17 (emphasis in the original).

5. The word *earworm* derives from the German word *ohrwurm* and first appeared in English in Desmond Bagley's 1978 novel *Flyaway*.

6. Sacks, *Musicophilia*, 41–48.

7. Wan et al., "The Therapeutic Effects of Singing in Neurological Disorders," 287. The authors acknowledge that although the ability to sing is not dependent on formal vocal training, it "can be enhanced by training."

8. Typically, the more you wish to be heard, the more you "raise" your voice. This raising is not limited to a higher volume, but also applies to a higher pitch. That's fine when you need to bark out safety instructions in a dangerous situation, but it's not necessarily best for singing. See Wertheim, "Is There an Anatomical Localisation for Musical Faculties?," 293, noting that it has long been recognized that the brain perceives higher frequency sounds more rapidly than it does lower frequency sounds.

9. A study by a group in Michigan measured increased levels of oxytocin and decreased levels of adrenocorticotrophic hormone (ACTH) in people before and after participating in group singing. The researchers attributed the decrease in ACTH, which leads to reduced cortisol levels, to stress reduction from singing. The increase in oxytocin levels was attributed to social bonding through singing as a group. This is consistent with the concept that entrainment, the ability to coordinate our motor movements to an external rhythmic stimulus, is pro-social: the members, entraining with one another through group singing, bonded with one another. See Keeler et al., "The Neurochemistry and Social Flow of Singing," 7–8.

10. Somayaji et al., "Acute Effects of Singing on Cardiovascular Biomarkers," 1.

11. Ronca, "Why do people sing in the shower?," https://science.howstuffworks.com/life/biology-fields/sing-in-the-shower.htm.

12. For more information, visit Victor Café, "History: Our Story." https://victorcafe.com/History-2.

13. I'm not the only one who's crazy about this song. Aznavour's fame was such that a show to preserve his legacy was created with his input shortly before his death. The title of the show? *Formidable!* with a subtitle of *The History of a Legend.* See, for example: https://themontrealeronline.com/2022/05/formidable-aznavour-the-history-of-a-legend/.

14. Schubert, *Listen: How Pete Seeger Got America Singing*, 3.

15. For more on this community organization of vocalists, visit Ridgefield Chorale, https://ridgefieldchorale.org/home/. Efforts to address the concern of Daniela Sikora and others—namely, that music is no longer part of

the curriculum in many schools—are being made. See, for example, *Music for Every Child* (https://sfcm.edu/sites/default/files/SFCM-Music_for_Every _Child.pdf), a white paper authored by Indre Viskontas, PhD. Please also read this book's afterword by Scott C. Shuler, PhD.

16. On this capability, see chapter 4.

17. Zentner and Eerola, "Rhythmic Engagement with Music in Infancy," 5768. Auditory-motor synchronization is the more formal term for entrain-ment. Brown, Martinez, and Parsons, "The Neural Basis of Human Dance," 1157, describe three fundamental components of dance: entrainment, me-ter (planned movements to a regular beat), and patterned movement (body movements organized into spatial patterns). Their research reveals that dif-ferent brain areas serve each component.

18. The pleasurable sensation of wanting to move to music's rhythm is known as groove. Vander Elst et al., "The Neuroscience of Dance," 18.

19. "From Wounds to Wisdom," *facebook.com*, https://www.facebook .com/watch/?v=3249884081754742.

20. I do not endorse dancing in the shower—it increases the risk of fall-ing to a level that's far too dangerous.

21. Harrison et al., "Graceful Gait," 5. This paper reports improved mo-bility and reduced fall risk among older (median age, late sixties) women participating in ballet classes online. Reduced fall risk was not detected in a control group participating in virtual wellness information sessions.

22. Loersch and Arbuckle, "Unraveling the Mystery of Music," 777.

23. Bailando Journey, "Why Does Dance Bring More Romance into Your Life?" https://bailandojourney.com/2021/02/25/why-does-dance-bring-more -romance-into-your-life/.

24. In French, the word *distraction* often conveys a sense of amuse-ment or entertainment.

25. For more, see Vitti's Dance Studio, https://www.vittisdancestudio .com/.

26. "The embodiment of emotions" is the technical term for expressing emotions through movement. Park et al., "Why Do Humans—and Some Animals—Love to Dance?," 5.

27. Scheer, "Rhythm Research and Sources," https://www.rhythmre-searchresources.net/research-drum-therapy-introduction.html. Admittedly being a bit of a drum wonk, Dave related that he has amassed a collection of drums, Western style as well as African style, over the years. He is particu-larly intrigued these days by djembe drums, from Western Africa. He ex-plained that the drums are crafted from the hollowed-out trunk of a single

tree. Traditionally, a stretched goat skin would serve as the head of the drum. African goat skins cannot, however, be imported into the United States, so drum heads made from cattle skin or, more commonly, plastic are attached here. In fact, most drum heads made in the United States at present are made of plastic.

28. As an historical note, pianists sometimes play trills (using two fingers, not two mallets, of course) despite pianos having a sustain pedal. This is a vestige from the time of the harpsichord, which does not have a sustain pedal. Pianos gradually replaced harpsichords for the most part, but pianists enjoy playing, and audiences enjoy hearing, trills for their artistic flourish. Consequently, trills never disappeared.

29. Harmony notes can be added since multiple bells can be rung simultaneously.

30. For more information about this community organization of bell and chime ringers, visit Chime In!, https://sites.google.com/chimeinmusic .org/chimein/home.

31. If you want to participate in a group in your area, an internet search or word-of-mouth should provide you with options. Examples include a choir or chorale (sing), exercise classes or formal studio (dance), drum circle or hand bell ensemble (play). You'll be immersing yourself in great music that you help to make!

32. Ruth and Müllensiefen, "Survival of Musical Activities," 8.

33. Kunst, "Share of Americans Who Played Musical Instruments in the Last 12 Months in 2018, by Age," https://www.statista.com/statistics /352204/number-of-people-play-musical-instrument-usa/.

34. Barry Mann and Cynthia Weil, *Make Your Own Kind of Music.* Screen Gems-EMI Music Inc, 1969.

Afterword

1. Howard Gardner, *Frames of Mind: The Theory of Multiple Intelligences* (New York: Basic Books, 2004), 99–127.

2. The Quadrivium comprises arithmetic, astronomy, geometry, and music.

3. Lelouda Stamou, "Plato and Aristotle on Music and Music Education: Lessons from Ancient Greece." *International Journal of Music Education.* 39 (2002): 3–16 (p. 5).

4. Or more concisely, as Brent Wilson—an esteemed visual arts colleague—dryly opined during a meeting where we were struggling to develop a consensus multi-arts framework for the 1994 National Assessment of

Educational Progress in the Arts: "You musicians go for the tingle." (personal memory).

5. Saint Augustine. "Confessions (10:33)." Translated by J. G. Pilkington. In *Basic Writings of Saint Augustine*, edited by Whitney J. Oates, 171–172. New York: Random House, 1948.

6. Newkey-Burden, Chas. "Why Chechnya has banned music that is 'too fast or too slow'." *The Week*, April 19, 2024, 9.

7. https://en.wikiquote.org/wiki/Andrew_Fletcher.

8. Langer, Susanne K. *Feeling and Form: A theory of art developed from Philosophy in a New Key*. New York: Charles Scribner's Sons, 1953, 27.

9. Mark, Michael and Charles Gary. *A History of American Music Education, 3rd Edition*, 160–162. Reston, VA: MENC: The National Association for Music Education, 2007.

10. "Neurologist Insists that Music, as Well as Life, Can Begin When One is 40." *New York Times*, July 13, 1986, Section 1, Page 39.

11. Personal recollection.

12. National Coalition for Core Arts Standards (2012). *National Core Arts Standards: A Conceptual Framework for Arts Learning*. https://www.nationalartsstandards.org/sites/default/files/Conceptual%20Framework%2007-21-16.pdf.

13. Meyer, Leonard B. *Emotion and Meaning in Music*. Chicago: The University of Chicago Press, 1974, 30–32.

Bibliography

Adachi, Mayumi, and Sandra Trehub. "Musical Lives of Infants." In *The Oxford Handbook of Music Education*, edited by Gary McPherson and James Welch, 229–47. Oxford University Press, 2012.

Allen, John. *The Lives of the Brain.* Belknap Press of Harvard University Press, 2009.

Altenmüller, Eckart, and Gottfried Schlaug. "Apollo's Gift: New Aspects of Neurologic Music Therapy." *Progress in Brain Research* 217 (2015): 237–52. https://doi.org/10.1016/bs.pbr.2014.11.029.

Ashoori, Aidin, David Eagleman, and Joseph Jankovic. "Effects of Auditory Rhythm and Music on Gait Disturbances in Parkinson's Disease." *Frontiers in Neurology* 6 (2015): article 234. https://doi.org/10.3389/fneur.2015.00234.

"Baby Finnley Gets Emotional While Hearing Classical Music for the First Time." YouTube, January 17, 2015. https://www.youtube.com/watch?v=WDP7liBCliQ.

"Baby Gets Emotional When Mom Sings Opera!" YouTube, July 27, 2018. https://www.youtube.com/watch?v=jME9QyhHS_0.

"Baby React [*sic*] to Music." YouTube, October 18, 2022. https://www.youtube.com/shorts/SJ9OnIxtEQE.

Bailando Journey. "Why Does Dance Bring More Romance into Your Life?" *bailandojourney.com*, February 25, 2021, https://bailandojourney.com/2021/02/25/why-does-dance-bring-more-romance-into-your-life/.

Baird, Amee, Heidi Umbach, and William Thompson. "A Nonmusician with Severe Alzheimer's Dementia Learns a New Song." *Neurocase* 23, no. 1 (February 2017): 36–40. https://doi.org/10.1080/13554794.2017.1287278.

Balbag, M. Alison, Nancy L. Pedersen, and Margaret Gatz. "Playing a Musical Instrument as a Protective Factor Against Dementia and Cognitive Impairment: A Population-Based Twin Study." *International Journal of Alzheimer's Disease.* 2014. https://doi.org/10.1155/2014/836748.

Battelli, Lorella, Vincent Walsh, Alvaro Pascual-Leone, and Patrick Cavanagh. "The 'When' Parietal Pathway Explored by Lesion Studies." *Current Opinion in Neurobiology* 18, no. 2 (2008): 120–26. https://doi.org/10.1016/j.conb.2008.08.004.

Beck, Guy. *Sonic Theology: Hinduism and Sacred Sound*. University of South Carolina Press, 1993.

Bekirska, Boryana. "Babies Listening for the First Time Classical Music." YouTube, February 25, 2015. https://www.youtube.com/watch?v=XZapawRtSnE.

Black, Riley. "Ancient Egyptian Love Songs Connect Romance Across Time." Natural History Museum of Utah, September 27, 2021. https://nhmu.utah.edu/articles/2023/05/ancient-egyptian-love-songs-connect-romance-across-time.

Bomzer, Ryan. "Johannes Brahms–Lullaby (Kalima Tab)." *Carved Culture*, January 26, 2023. https://www.carvedculture.com/blogs/articles/johannes-brahms-lullaby-kalimba-tab.

Brandt, Thomas, and Doreen Huppert. "Brain Beats Heart: A Cross-Cultural Reflection." *Brain*. 144, no. 6 (2021):1617–20. https://doi.org/10.1093/brain/awab080.

Bratu, Ionut-Flavius, Adriana Nica, Irina Oane, Andrei Daneasa, Sergiu Stoica, Andrei Barborica, and Ioana Mindruta. "Musicogenic Seizures in Temporal Lobe Epilepsy: Case Reports Based on Ictal Source Localization Analysis." *Frontiers in Neurology* 14 (2023): 1072075. https://doi.org/10.3389/fneur.2023.1072075.

Brenowitz, Eliot, and Tracy Larson. "Neurogenesis in the Adult Avian Song-Control System." *Cold Spring Harbor Perspectives in Biology* 7, no. 6 (2015): a019000. https://doi.org/10.1101/cshperspect.a019000.

Brown, Steven, Michael Martinez, and Lawrence Parsons. "Music and Language Side by Side in the Brain: A PET Study of the Generation of Melodies and Sentences." *European Journal of Neuroscience* 23, no. 10 (2006): 2791–2803. https://doi.org/10.1111/j.1460-9568.2006.04785.x.

Brown, Steven, Michael Martinez, and Lawrence Parsons. "The Neural Basis of Human Dance." *Cerebral Cortex* 16, no. 8 (2006): 1157–67. https://doi.org/10.1093/cercor/bhj057.

Brust, John. "Music and the Neurologist: A Historical Perspective" *Annals of the New York Academy of Sciences* 930, no. 1 (2006): 143–52. https://doi.org/10.1111/j.1749-6632.2001.tb05730.x.

Central Michigan University, "Music Therapy Helps Elderly with Alzheimer's and Dementia Recall Long-Term Memories." https://www.youtube.com/watch?v=aKQsxKhtDos.

Cepelewicz, Jordana. "In Birds' Songs, Brains and Genes, He Finds Clues to Speech." *Quanta Magazine*, January 30, 2018. https://www.quantamagazine.org/in-birds-songs-brains-and-genes-he-finds-clues-to-speech-20180130/.

Chatterjee, Anjan. *The Aesthetic Brain: How We Evolved to Desire Beauty and Enjoy Art*. Oxford University Press, 2014.

Chea, Maryane, Amina Ben Salah, Monica Toba, Ryan Zeineldin, Brigitte Kaufmann, Agnès Weill-Chounlamountry, Lionel Naccache, Eléonore Bayen, Paolo Bartolomeo. "Listening to Classical Music Influences Brain Connectivity in Post-Stroke Aphasia: A Pilot Study," *Annals of Physical and Rehabilitation Medicine* 67, no. 4 (2024): 101825. doi: 10.1016/j.rehab.2024.101825. Epub 2024 Mar 12. PMID: 38479248.

Chen, Joyce, Virginia Penhune, and Robert Zatorre. "Listening to Musical Rhythms Recruits Motor Regions of the Brain." *Cerebral Cortex* 18, no. 12 (2008): 2844–54. https://doi.org/10.1093/cercor/bhr042.

christinahoang76. "Baby's Response to Horrible Singing." YouTube, May 10, 2013, https://www.youtube.com/shorts/NJpuwxuAOdY.

Cognitive Neuroscience Society. "From Lullabies to Live Concerts: How Music and Rhythm Shape Our Social Brains." *ScienceDaily*, press release, March 27, 2018. www.sciencedaily.com/releases/2018/03/180327102835.htm.

Cohen, Robert S. "Alzheimer's Stories (2010) (for soloists, chorus, and 14 players)." *wisemusicclassical.com*. (No date indicated): https://www.wisemusicclassical.com/work/64118/Alzheimers-Stories--Robert-S-Cohen/.

Collins, Lorence. "Does the Bible Contradict Accepted Biological Concepts?" https://scholarworks.calstate.edu/downloads/mw22v9266. Originally published in *Creation/Evolution* 36 (Summer 1995): 15–23.

"Consonant vs Dissonant Intervals: Ear Training." YouTube, April 29, 2013. https://www.youtube.com/watch?v=fabeLu4-Ja0.

Cooksey, John. "4 Month Old Baby Tries to Sing to Karen Carpenter Song, Melts Hearts." YouTube, June 9, 2014. https://www.youtube.com/watch?v=bEeizaWjdXw.

Copland, Aaron. *What to Listen for in Music*. Penguin, 1953.

Critchley, MacDonald, and R. A. Henson. *Music and the Brain: Studies in the Neurology of Music*. William Heinemann Medical Books, 1977.

Curry, Andrew. "Imaging the Past: What Stone Toolmaking and Neuroscience Can Tell Us About What It Means to Be Human." *Archaeology* 71, no. 2 (2018): 42–45.

Dalla Bella, Simone, and Jakub Sowinski. "Uncovering Beat Deafness: Detecting Rhythm Disorders with Synchronized Finger Tapping and Perceptual Timing Tasks." *Journal of Visualized Experiments* 97 (March 2015): e51761. https://doi.org/10.3791/51761-v.

Daly, David, and Timothy Pedley. *Current Practice of Clinical Electroencephalography*. Raven Press, 1990.

Damasio, Antonio. *Self Comes to Mind*. Vintage, 2012.

Damasio, Antonio. *Looking for Spinoza: Joy, Sorrow, and the Feeling Brain*. Mariner, 2003.

Damasio, Antonio. *Descartes' Error*. Penguin, 1994.

Davies, Nina. *Ancient Egyptian Paintings*. University of Chicago Press, 1936.

Davis, William, Kate Gfeller, and Michael Thaut. *An Introduction to Music Therapy Theory and Practice*. American Music Therapy Association, 2008.

Dennett, Daniel. "Review of *Descartes' Error* by Antonio Damasio." *Times Literary Supplement*, August 25, 1995.

DeVale, Sue Carole, and Bo Lawergren. "Harp: IV. Asia." Grove Music Online. Oxford University Press. https://doi.org/10.1093/gmo/9781561592630 .article.45738.

Devinsky, Orrin, Martha Morrell, and Brent Vogt. "Contributions of Anterior Cingulate Cortex to Behaviour." *Brain* 118 (1995): 279–306. https://doi.org/10.1093/brain/118.1.279.

Dinneen, James. "Male and Female Gibbons Sing Duets in Time with Each Other." *NewScientist*, January 11, 2023. https://www.newscientist.com /article/2354278-male-and-female-gibbons-sing-duets-in-time-with -each-other/.

"Donkey Riding—with Capstan Demonstration." YouTube, June 26, 2009. https://www.youtube.com/watch?v=XJpVXZVlklM&list=PL53A91BE 7CAF6DB5B&index=2.

Dubuc, Bruno, Patrick Robert, Denis Paquet, and Al Daigen. "Dusting Off the Triune Brain and the Limbic System." *The Brain from Top to Bottom* (blog), October 29, 2012. https://www.blog-thebrain.org/blog/2012/10/29/dusting-off-the-triune-brain-and-the-limbic-system/.

Eichler, Jeremy. *Time's Echo: The Second World War, the Holocaust, and the Music of Remembrance.* Knopf Doubleday, 2023.

"Eight Oldest Musical Instruments in the World." *Oldest.org*, 2023. https://www.oldest.org/music/musical-instruments/.

El Haj, Mohamad, Luciano Fasotti, and Philippe Allain. "The Involuntary Nature of Music-Evoked Autobiographical Memories in Alzheimer's Disease." *Consciousness and Cognition* 21 (2012): 238–46. https://doi.org/10.1016/j.concog.2011.12.005.

Fang, Rong, Shengxuan Ye, Jiangtao Huangfu, and David Calimag. "Music Therapy Is a Potential Intervention for Cognition of Alzheimer's Disease: A Mini-Review." *Translational Neurodegeneration* 6, no. 2 (2017). https://doi.org/10.1186/s40035-017-0073-9.

Ferreri, Laura, Ernest Mas-Herrero, Robert J. Zatorre, and Antoni Rodriguez-Fornells. "Dopamine Modulates the Reward Experiences Elicited by Music." *Proceedings of the National Academy of Sciences* 116, no. 9 (2019): 3793–98. https://doi.org/10.1073/pnas.1811878116.

Fitch, W. Tecumseh. "The Evolution of Music in Comparative Perspective." *Annals of the New York Academy of Sciences* 1060 (2005): 1–21. https://doi.org/10.1196/annals.1360.004.

Fleming, Renée, ed. *Music and Mind: Harnessing the Arts for Health and Wellness.* Viking, 2024.

"4 Month Old Baby Tries to Sing to Karen Carpenter Song, Melts Hearts." YouTube, June 9, 2014. https://www.youtube.com/watch?v=bEeizaWjdXw.

"From Wounds to Wisdom." September 7, 2020, https://www.facebook.com/watch/?v=3249884081754742.

"Gait Training for Parkinson's Patient Using Music." https://www.youtube.com/watch?v=uDjQ7lKmH3s.

Garrido-Vásquez, Patricia, Marc Pell, Silke Paulmann, and Sonja Kotz. "Dynamic Facial Expressions Prime the Processing of Emotional Prosody." *Frontiers in Human Neuroscience* 12 (2018): article 244. https://doi.org/10.3389/fnhum.2018.00244.

Gault, Robert. "Touch as a Substitute for Hearing in the Interpretation and Control of Speech." *JAMA Otolaryngology Head Neck Surgery* 3, no. 2 (1926): 121–35.

Gazzaniga, Michael. *Who's in Charge?* HarperCollins, 2011.

Geissmann, Thomas. "Gibbon Songs and Human Music from an Evolutionary Perspective." In *The Origins of Music,* edited by Nils Wallin, Björn Merker, and Steven Brown, 103–23. MIT Press, 2000.

Geschwind, Michael. "Are You Related to 'the Geschwind?'" *Neuropsychology Review* 20, no. 2 (2010): 123–25. https://doi.org/10.1007/s11065-010-9135-9.

Geschwind, Norman. "Neurological Knowledge and Complex Behaviors." *Cognitive Science* 4 (1980): 185–93. https://doi.org/10.1207/s15516709cog0402_3.

Get America Singing . . . Again. Foreword by Pete Seeger. 2 vols. Hal Leonard Corporation, 1996.

Gold, Benjamin, Ernest Mas-Herrero, Yashar Zeighami, Mitchel Benovoy, Alain Dagher, and Robert Zatorre. "Musical Reward Prediction Errors Engage the Nucleus Accumbens and Motivate Learning." *Proceedings of the National Academy of Sciences* 116, no. 8 (2019): 3310–15. https://doi.org/10.1073/pnas.1809855116.

Gold, Benjamin, Marcus Pearce, Ernest Mas-Herrero, Alain Dagher, and Robert Zatorre. "Predictability and Uncertainty in the Pleasure of Music: A Reward for Learning?" *Journal of Neuroscience* 39, no. 47 (2019): 9397–409. https://doi.org/10.1523/JNEUROSCI.0428-19.2019.

Goodrich, Joanna. "How This Record Company Engineer Invented the CT Scanner." *IIE Spectrum,* January 12, 2023. https://spectrum.ieee.org/invention-of-ct-scanner.

Gosselin, Nathalie, Isabelle Peretz, Erica Johnsen, and Ralph Adolphs. "Amygdala Damage Impairs Emotion Recognition from Music." *Neuropsychologia* 45, no. 2 (2007): 236–44. https://doi.org/10.1016/j.neuropsychologia.2006.07.012.

Gould, Evlyn. *The Fate of Carmen.* Johns Hopkins University Press, 1996.

Grau-Sánchez, Jenny, E. Duarte, N. Ramos-Escobar, J. Sierpowska, N. Rueda, S. Redon, M. Veciana de las Heras, J. Pedro, T. Särkämö, and A. Rodriguez-Fornells. "Music-Supported Therapy in the Rehabilitation of Subacute Stroke Patients: A Randomized Controlled Trial." *Annals of Physical and Rehabilitation Medicine* 61, supp. (2018): e191. https://doi.org/10.1016/j.rehab.2018.05.438.

Grieser, DiAnne, and Patricia Kuhl. "Maternal Speech to Infants in a Tonal Language: Support for Universal Prosodic Features in Motherese."

Developmental Psychology 24, no. 1 (1988): 14–20. https://doi.org/10
.1037/0012-1649.24.1.14.

Groussard, Mathilde, Fausto Viader, Brigitte Landeau, Béatrice Desgranges,
Francis Eustache, and Hervé Platel. "The Effects of Musical Practice on
Structural Plasticity: The Dynamics of Grey Matter Changes." *Brain and
Cognition* 90 (2014): 174–80. https://doi.org/10.1016/j.bandc.2014.06
.013.

Gunlock, Anicea. "Gait Training for Parkinsons's [sic] Patient Using Music."
YouTube, January 13, 2017. https://www.youtube.com/watch?v=uDj
Q7lKmH3s.

Guzzo, Carla Mesa. "Hiddenness and Darkness in Ancient Egyptian Love
Songs. *Nile Scribes*, February 16, 2019. https://nilescribes.org/2019/02/16
/egyptian-love-songs/.

Hanna-Pladdy, Brenda, and Alicia MacKay. "The Relation Between Instrumental Music Activity and Cognitive Aging." *Neuropsychology* 25, no. 3
(2011): 378–86.

Harding, Cheryl. "Learning from Bird Brains: How the Study of Songbird
Brains Revolutionized Neuroscience." *Lab Animal* 33, no. 5 (2004): 28–
33. https://doi.org/10.1038/laban0504-28.

Harlow, Frederick Pease. *Chanteying Aboard American Ships*. Mystic Seaport
Museum, 2004.

Harrison, Elinor, Allison Haussler, Lauren Tueth, Sidney Baudendistel, and
Gammon Earhart. "Graceful Gait: Virtual Ballet Classes Improve Mobility and Reduce Falls More than Wellness Classes for Older Women."
Frontiers in Aging Neuroscience 16 (2024): 1289368. https://doi.org/10
.3389/fnagi.2024.1289368.

Haslbeck, Friederike, and Dirk Bassler. "Music from the Very Beginning: A
Neuroscience-Based Framework for Music as Therapy for Preterm Infants and Their Parents." *Frontiers in Behavioral Neuroscience* 12 (2018).
https://doi.org/10.3389/fnbeh.2018.00112.

Hebb, Donald. *The Organization of Behavior: A Neuropsychological Theory*.
John Wiley & Sons, 1949.

Helmuth, Laura. "Why Bird Brains Bloom in Spring." *Smithsonian Magazine*,
March 7, 2011. https://www.smithsonianmag.com/science-nature/why
-bird-brains-bloom-in-spring-45154088/.

Hepper, Peter. "An Examination of Fetal Learning Before and After Birth." *Irish Journal of Psychology* 12 (1991): 95–107. https://doi.org/10.1080/03033910.1991.10557830.

The Holy Bible, Revised Standard Version, second ed. Edited by Jeff Cavins et al. Oxford University Press, 1963.

Homick, David. *From Time to Time: A Time Travel Romantic Thriller.* Farmington, NY: Blue Knight Media, 2018.

Idaho Fish and Game. "Dance of the Sage Grouse." YouTube, April 10, 2018. https://www.youtube.com/watch?v=28D491RoBEs.

imazimbo2. "Consonant vs Dissonant Intervals—Ear Training." *YouTube.com*, April 29, 2013. https://www.youtube.com/watch?v=fabeLu4-Ja0.

Inman, Cory, Kelly Bijanki, David Bass, Robert Gross, Stephan Hamann, and Jon Willie. "Human Amygdala Stimulation Effects on Emotion Physiology and Emotional Experience." *Neuropsychologia* 145 (2020): 106722. https://doi.org/10.1016/j.neuropsychologia.2018.03.019.

Insight & Ammunition. "Baby Finnley Gets Emotional While Hearing Classical Music for the First Time." YouTube, January 17, 2015. https://www.youtube.com/watch?v=WDP7IiBCliQ.

Institut Marquès. "Music Improves In-Vitro Fertilization." *institutomarques.com*, n.d. https://institutomarques.com/reproduccion-asistida/musica-y-fecundacion/?_gl=1*lxzk95*_up*MQ..*_ga*NTk1MjM1MDgwLjE3NDE4NDExNjc.*_ga_EPEV9MTWFX*MTc0MTg0MTE2NS4xLjEuMTc0MTg0MTIyNy4wLjAuMjA2OTAzOTkzNw.

"An Introduction to the Life and Work of John Hughlings Jackson," *Medical History Supplement* 26 (2007): 3–34.

Jacobsen, Jörn-Henrik, Johannes Stelzer, Thomas Hans Fritz, Gael Chételat, Renaud La Joie, and Robert Turner. "Why Musical Memory Can Be Preserved in Advanced Alzheimer's Disease." *Brain* 138, no. 8 (2015): 2438–50. https://doi.org/10.1093/brain/awv135.

Johnson, Julene, and Amy Graziano. "Some Early Cases of Aphasia and the Capacity to Sing." *Progress in Brain Research* 216 (2015): 73–89. https://doi.org/10.1016/bs.pbr.2014.11.004.

Jolij, Jacob, and Maaike Meurs. "Music Alters Visual Perception." *PLoS ONE* 6, no. 4 (2011): e18861. https://doi.org/10.1371/journal.pone.0018861.

Jordan, David K. "Four Ancient Chinese Songs," March 31, 2024. https://pages.ucsd.edu/~dkjordan/chin/chtxts/ShyJing.html.

Kahneman Daniel. *Thinking, Fast and Slow*. Farrar, Straus, and Giroux, 2011.

Kandel, Eric. *The Age of Insight*. Random House, 2012.

Kaneda, Mizuki, and Naoyuki Osaka. "Role of Anterior Cingulate Cortex During Semantic Coding in Verbal Working Memory." *Neuroscience Letters* 436 (2008): 57–61. https://doi.org/10.1016/j.neulet.2008.02.069.

Karkou, Vicky and Bonnie Meekums. "Dance Movement Therapy for Dementia." *Cochrane Database of Systematic Reviews* 2 (February 2017): CD011022. https://doi.org/10.1002/14651858.CD011022.pub2.

Katzman, Robert. "Education and the Prevalence of Dementia and Alzheimer's Disease." *Neurology* 43 (1993): 13–20. https://doi.org/10.1212/wnl.43.1_part_1.13.

Katzman, Robert. "The Prevalence and Malignancy of Alzheimer Disease: A Major Killer." *Archives of Neurology* 33, no. 4 (1976): 217–18. https://doi.org/10.1001/archneur.1976.00500040001001-01.

Keeler, Jason, Edward Roth, Brittany Neuser, John Spitsbergen, Daniel Waters, and John-Mary Vianney. "The Neurochemistry and Social Flow of Singing: Bonding and Oxytocin." *Frontiers in Human Neuroscience* 9 (2015): article 518. https://doi.org/10.3389/fnhum.2015.00518.

Klein, Joanna. "Some Songbirds Have Brains Specially Designed to Find Mates for Life." *New York Times*, February 13, 2018.

Klimova, Blanka, Martin Valis, and Kamil Kuca. "Cognitive Decline in Normal Aging and Its Prevention: A Review on Non-Pharmacological Lifestyle Strategies." *Clinical Interventions in Aging* 12 (2017): 903–10. https://doi.org/10.2147/CIA.S132963.

Konnikova, Maria. "The Man Who Couldn't Speak and How He Revolutionized Psychology." *Scientific America* (blog), February 8, 2013. https://blogs.scientificamerican.com/literally-psyched/the-man-who-couldnt-speakand-how-he-revolutionized-psychology/.

Konoike, Naho, and Katsuki Nakamura. "Cerebral Substrates for Controlling Rhythmic Movements." *Brain Science* 10, no. 8 (2020): 514. https://doi.org/10.3390/brainsci10080514.

Konoike, Naho, Yuka Kotozaki, Hyeonjeong Jeong, Atsuko Miyazaki, Kohei Sakaki, Takamitsu Shinada, Mtotaki Sugiura, Ryuta Kawashima, and Katsuki Nakamura. "Temporal and Motor Representation of Rhythm in Fronto-Parietal Cortical Areas: An fMRI Study." *PLoS ONE* 10. no. 6 (2015): e0130120. https://doi.org/10.1371/journal.pone.0130120.

Kraus, Nina. *Of Sound Mind: How Our Brain Constructs a Meaningful Sonic World.* MIT Press, 2021.

Kung, Shu-Jen, Joyce Chen, Robert Zatorre, and Virginia Penhune. "Interacting Cortical and Basal Ganglia Networks Underlying Finding and Tapping to the Musical Beat." *Journal of Cognitive Neurosciences* 25 no. 3 (2012): 401–20. https://doi.org/10.1162/jocn_a_00325.

Kunst, Alexander. "Share of Americans Who Played Musical Instruments in the Last 12 Months in 2018, by Age." *Statista*. https://www.statista.com /statistics/352204/number-of-people-play-musical-instrument-usa/.

Launay, Jacques, and Eiluned Pearce. "Choir Singing Improves Health, Happiness—and Is the Perfect Icebreaker." *Conversation*, October 28, 2015. https://theconversation.com/choir-singing-improves-health-happiness -and-is-the-perfect-icebreaker-47619.

Levitin, Daniel. *The World in Six Songs: How the Musical Brain Created Human Nature*. Dutton, 2016.

Levitin, Daniel. *This Is Your Brain on Music: The Science of a Human Obsession*. Dutton, 2006.

Li, Danny. "How We Process Timbre (Tone)." Presented at Neuroscience for Musicians, sponsored by the Johns Hopkins Center for Music and Medicine, June 29, 2021. https://www.youtube.com/watch?v =ipJo7Z1P4Ug.

Library of Congress. "African American Spirituals." *loc.gov* (n.d): https://www .loc.gov/item/ihas.200197495/.

Loersch, Chris, and Nathan Arbuckle. "Unraveling the Mystery of Music: Music as an Evolved Group Process." *Journal of Personality and Social Psychology* 105, no. 5 (2013): 777–98.

Loui, Psyche, Catherine Wan, and Gottfried Schlaug. "Neurological Bases of Musical Disorders and Their Implications for Stroke Recovery." *Acoustics Today* 6, no. 3 (2010): 28–36. https://doi.org/10.1121/1.3488666.

Lucas, John. "Memory, Overview." In *Encyclopedia of the Human Brain*, 817–33. Academic Press, 2002. https://publications.ascendensasia.com /encyclopedia-of-human-brain-by-antonio-rdamasio-martha-jfarah -michael-fhuerta-vs-ramachandran/68491824.

Macmillan, David. "A Self-Guided Walk on the Artists' Trail in Montmartre." Montmartre Artists' Studios, 2025. https://montmartrefootsteps.com /montmartre-historical-cultural-context/.

Maddock, R. J., A. S. Garrett, and M. H. Buonocore. "Remembering Familiar People: The Posterior Cingulate Cortex and Autobiographical Memory Retrieval." *Neuroscience* 104, no. 3 (2001): 667–76. https://doi.org/10.1016/S0306-4522(01)00108-7/.

Mainka, Stefan, Jörg Wissel, Heinz Völler, and Stefan Evers. "The Use of Rhythmic Auditory Stimulation to Optimize Treadmill Training for Stroke Patients: A Randomized Controlled Trial." *Frontiers in Neurology* 9 (2018): article 755. https://doi.org10.3389/fneur.2018.00755.

Mann, Barry, and Cynthia Weil. *Make Your Own Kind of Music.* Screen Gems-EMI Music Inc, 1969.

Marshall, Penny, dir. *Awakenings.* Lasker/Parkes Productions, 1990.

Martínez-Molina, Noelia, Ernest Mas-Herrero, Antoni Rodriguez-Fornells, Robert Zatorre, and Josep Marco-Pallares. "Neural Correlates of Specific Musical Anhedonia." *Proceedings of the National Academy of Sciences* 113 (2016): E7337–E7345. https://doi.org/10.1073/pnas.16112111.

Martínez-Molina, Noelia, Ernest Mas-Herrero, Antoni Rodriguez-Fornells, Robert Zatorre, and Josep Marco-Pallares. "White Matter Microstructure Reflects Individual Differences in Music Reward Sensitivity." *Journal of Neuroscience* 39, no. 25 (2019):5018–27. https://doi.org/10.1523/JNEUROSCI.2020-18.2019.

Martins, Ralph. "Was 'Earliest Musical Instrument' Just a Chewed-Up Bone?" *National Geographic*, March 31, 2015. https://www.nationalgeographic.com/science/article/150331-neanderthals-music-oldest-instrument-bones-flutes-archaeology-science.

Masataka, Nobuo. "Motherese in a Signed Language." *Infant Behavior and Development* 15, no. 4 (1992): 453–60. https://doi.org/10.1016/0163-6383(92)80013-K.

Masataka, Nobuo. "Preference for Consonance over Dissonance by Hearing Newborns of Deaf Parents and of Hearing Parents." *Developmental Science* 9, no. 1 (2006): 46–50. https://doi.org/10.1111/j.1467-7687.2005.00462.x.

Matziorinis, Anna Maria, and Stefan Koelsch. "The Promise of Music Therapy for Alzheimer's Disease: A Review." *Annals of the New York Academy of Sciences* 1516, no. 1 (2022): 11–17. https://doi.org/10.1111/nyas.14864.

McChesney-Atkins, Susan, Keith Davies, Georgia Montouris, John Silver, and Daniel Menkes. "Amusia after Right Frontal Resection for Epilepsy with

Singing Seizures: Case Report and Review of the Literature." *Epilepsy & Behavior* 4 (2003): 343–47. https://doi.org/ 10.1016/s1525-5050(03)00079-9.

McQuay, Bill, and Christopher Joyce. "How Sound Shaped the Evolution of Your Brain." *Morning Edition*, NPR, September 10, 2015. https://www.npr.org/sections/health-shots/2015/09/10/436342537/how-sound-shaped-the-evolution-of-your-brain.

MedRhythms. "Melodic Intonation Therapy—Longer Phrases." https://www.youtube.com/watch?v=t0Vv-JHhIsl.

Mehr, Samuel, Manvir Singh, Hunter York, Luke Glowacki, and Max Krasnow. "Form and Function in Human Song." *Current Biology* 28 (2018): 356–68. https://doi.org/10.1016/j.cub.2017.12.042.

Menke, Barbara, Joachim Hass, Carston Diener, and Johannes Pöschl. "Family-Centered Music Therapy—Empowering Premature Infants and Their Primary Caregivers Through Music: Results of a Pilot Study." *PLoS ONE* 16, no. 5 (2021): e0250071. https://doi.org/10.1371/journal.pone.0250071.

Molnar-Szakacs, Istvan, and Katie Overy. "Music and Mirror Neurons: From Motion to 'E'motion." *Social Cognitive and Affective Neuroscience* 1 (2006): 235–41. https://doi.org/10.1093/scan/nsl029.

Moore, Jordan, and Sarah Woolley. "Emergent Tuning for Learned Vocalizations in Auditory Cortex." *Nature Neuroscience* 22 (2019): 1469–76. https://doi.org/10.1038/s41593-019-0458-4.

Moreira, Shirlene, Francis Ricardo dos Reis Justi, and Marcos Moreira. "Can Musical Intervention Improve Memory in Alzheimer's Patients? Evidence from a Systematic Review." *Dementia & Neuropsychologia* 12, no. 2 (2018): 133–42. https://doi.org/10.1590/1980-57642018dn12-020005.

Mudcat Café. "Origin: If You're Happy and You Know It–How Old?" various dates. https://mudcat.org/thread.cfm?threadid=20648.

Naccache, Lionel. *Quatre exercices de pensée juive pour cerveaux réfléchis: Le judaïsme à la lumière des neurosciences*. Éditions InPress, 2003.

Naccache, Lionel. *Apologie de la discrétion*. Éditions Odile Jacob, 2022.

Nash, Alanna. "Folk-Pop Legend Gordon Lightfoot Goes Solo." AARP, March 27, 2020. https://www.aarp.org/entertainment/music/info-2020/gordon-lightfoot.html.

National Museum of Slovenia. "Neanderthal Flute: The Oldest Musical Instrument in the World," n.d. https://www.nms.si/en/collections/highlights/343-Neanderthal-flute.

Navarro, Laura, Fredrico Martinón-Torres, and Antonio Salas. "Sensogenomics and the Biological Background Underlying Musical Stimuli: Perspectives for a New Era of Musical Research." *Genes* 12 (2021): 1454. https://doi.org/10.3390/genes12091454.

Nissim, Nicole, Andrew O'Shea, Vaughn Bryant, Eric Porges, Ronald Cohen, and Adam Woods. "Frontal Structural Neural Correlates of Working Memory Performance in Older Adults." *Frontiers in Aging Neuroscience* 8 (2017): article 328. https://doi.org/10.3389/fnagi.2016.00328.

Ohbayashi, Machiko. "The Roles of the Cortical Motor Areas in Sequential Movements." *Frontiers in Behavioral Neuroscience* 15 (2021): article 640659. https://doi.org/ 10.3389/fnbeh.2021.640659.

Olds, James, and Peter Milner "Positive Reinforcement Produced by Electrical Stimulation of Septal Area and Other Regions of the Rat Brain." *Journal of Comparative Psychology* 47, no. 6 (1954): 419–27. https://doi.org/10.1037/h0058775.

Pandya, Sunil. "Understanding Brain, Mind and Soul: Contributions from Neurology and Neurosurgery." *Mens Sana Monographs* 9, no. (2011):129–49. PMID: 21694966; PMCID: PMC3115284. https://doi.org/10.4103/0973-1229.77431.

Panter, Barry, ed. *Creativity & Madness: Psychological Studies of Art and Artists.* AIMED Press, 2021.

Park, K. Shin, Madeleine Hackney, Christina Hugenschmidt, Christina Soriano, and Jennifer Etnier. "Why Do Humans—and Some Animals—Love to Dance?" *Frontiers for Young Minds* 10 (2022): 806631. https://doi.org/10.3389/frym.2022.806631.

Patel, Aniruddh. *Music, Language, and the Brain.* Oxford University Press, 2008.

Pearce, J.M.S. "Rufus of Ephesus (AD c. 80–150)." *Advances in Clinical Neuroscience and Rehabilitation*, June 20, 2018. https://acnr.co.uk/articles/rufus-of-ephesus/.

Penfield, Wilder, and Herbert Jasper. *Epilepsy and the Functional Anatomy of the Human Brain.* Little, Brown & Co, 1954.

Penhune, Virginia, and Robert Zatorre. "Rhythm and Time in the Premotor Cortex." *PLoS Biology* 17, no. 6 (2019): e3000293. https://doi.org/10.1371/journal.pbio.3000293.

Peretz, Isabelle. "The Biological Foundations of Music: Insights from Congenital Amusia." In *The Psychology of Music*, edited by Diana Deutsch, 551–64. Academic Press, 2013.

Peretz, Isabelle. *How Music Sculpts Our Brain*. Éditions Odile Jacob, 2020.

Peretz, Isabelle, and Robert Zatorre, eds. *The Cognitive Neuroscience of Music*. Oxford University Press, 2003.

Perrone-Capano, Carla, Floriana Volpicelli, and Umberto di Porzio. "Biological Bases of Human Musicality." *Reviews in the Neurosciences* 28, no. 3 (2017):235–45. https://doi.org/10.1515/revneuro-2016-0046.

Pinker, Steven, *How the Mind Works*. W. W. Norton, 1997.

Platel, Hervé, Jean-Claude Baron, Béatrice Desgranges, Frédéric Bernard, and Francis Eustache. "Semantic and Episodic Memory of Music Are Subserved by Distinct Neural Networks." *NeuroImage* 20, no. 1 (2003): 244–56. https://doi.org/10.1016/s1053-8119(03)00287-8.

Platel, Hervé, and Mathilde Groussard. "Benefits and Limits of Musical Interventions in Pathological Aging." In *Music and the Aging Brain*, edited by L. L. Cuddy, S. Belleville, and A. Moussard, 317–32. Elsevier Academic Press, 2020. https://doi.org/10.1016/B978-0-12-817422-7.00012-2.

"Premotor cortex." *Physiopedia*, June 12, 2024. https://www.physio-pedia.com/index.php?title=Premotor_Cortex&oldid=355222.

Reader's Digest. "Newborn Baby Mimics Dad's Voice and Makes Him Giggle." YouTube, April 23, 2023. https://www.youtube.com/shorts/FL1LUiqNITo.

Riding, Alan. "At 82, Charles Aznavour Is Singing a Farewell That Could Last for Years." *New York Times*, September 18, 2006. https://www.nytimes.com/2006/09/18/arts/music/at-82-charles-aznavour-is-singing-a-farewell-that-could-last-for.html.

Rogenmoser, Lars, Julius Kernbach, Gottfried Schlaug, and Christian Gaser. "Keeping Brains Young with Making Music." *Brain Structure and Function* 223 (2018): 297–305. https://doi.org/10.1007/s00429-017-1491-2.

Rolls, Edmund, Wei Cheng, and Jianfeng Feng. "The Orbitofrontal Cortex: Reward, Emotion, and Depression." *Brain Communications* 2, no. 2 (2020): fcaa196. https://doi.org/10.1093/braincomms/fcaa196.

Ronca, Debra. "Why Do People Sing in the Shower." *HowStuffWorks*, April 16, 2024. https://science.howstuffworks.com/life/biology-fields/sing-in-the-shower.htm.

Rosslau, Ken, Daniel Steinwede, C. Schröder, Sibylle Herholz, Claudia Lappe, Christian Dobel, and Eckart Altenmüller. "Clinical Investigations of Receptive and Expressive Musical Functions After Stroke." *Frontiers in Psychology* 6, no. 768 (2015): 1–11. https://doi.org/10.3389/fpsyg.2015.00768.

Ruiz-Muelle, Alicia, and María Mar López-Rodríguez. "Dance for People with Alzheimer's Disease: A Systematic Review." *Current Alzheimer Research* 16, no. 10 (2019): 919–33. https://doi.org/10.2174/1567205016666 190725151614.

Ruth, Nicolas, and Daniel Müllensiefen. "Survival of Musical Activities: When Do Young People Stop Making Music?" *PLoS ONE* 16(11) (2021): e0259105. https://doi.org/10.1371/journal.pone.0259105.

Sacks, Oliver. Foreword to *The Paradoxical Brain*, edited by Narinder Kapur, xiv–xv. Cambridge University Press, 2011.

Sacks, Oliver. *Musicophilia: Tales of Music and the Brain*. Knopf, 2007.

Saint-Georges, Catherine, Mohamed Chetouani, Raquel Cassel, Fabio Apicella, Ammar Mahdhaoui, Filippo Muratori, Marie-Christine Laznik, and David Cohen. "Motherese in Interaction: At the Cross-Road of Emotion and Cognition? (A Systematic Review)." *PLoS ONE* 8, no. 10 (2013): e78103. https://doi.org/10.1371/journal.pone.0078103.

Salakka, I, Anni Pitkäniemi, Emmi Pentikäinen, Kari Mikkonen, Pasi Saari, Petri Toiviainen, and Teppo Särkämö. "What Makes Music Memorable? Relationships Between Acoustic Musical Features and Music-Evoked Emotions and Memories in Older Adults." *PLoS ONE* 16, no. 5 (2021): e0251692. https://doi.org/10.1371/journal.pone.0251692.

Sankaran, Narayan, Matthew Leonard, Frederic Theunissen, and Edward Chang. "Encoding of Melody in the Human Auditory Cortex." *Science Advances* 10, no. 7 (2024): eadk0010. https://doi.org/10.1126/sciadv.adk0010.

Särkämö, Teppo, Mari Tervaniemi, and Minna Huotilainen. "Music Perception and Cognition: Development, Neural Basis, and Rehabilitative Use of Music." *Wiley Interdisplinary Reviews Cognitive Science* 4, no. 4 (2013): 441–51. https://doi.org/10.1002/wcs.1237.

Särkämö, Teppo, Mari Tervaniemi, Sari Laitinen, Ava Numminen, Merja Kurki, Julene Johnson, and Pekka Rantanen. "Cognitive, Emotional, and Social Benefits of Regular Musical Activities in Early Dementia: Randomized Controlled Study." *The Gerontologist* 54, no. 4 (2014): 634–50. https://doi.org/10.1093/geront/gnt100.

Schaal, Nora, Amir-Homayoun Javadi, Andrea Halpern, Bettina Pollok, and Michael Banissy. "Right Parietal Cortex Mediates Recognition Memory for Melodies." *European Journal of Neuroscience* 42, no. 1 (2015): 1660–66. https://doi.org/10.1111/ejn.12943.

Schacter, Daniel, Donna Rose Addis, Demis Hassabis, Victoria Martin, R. Nathan Spreng, and Karl Szpunar. "The Future of Memory: Remembering, Imagining, and the Brain." *Neuron* 76, no. 4 (2012): 677–94. https://doi.org/10.1016/j.neuron.2012.11.001.

Scheer, Peter. "Rhythm Research and Resources." *rhythmresearchresources.net*, June 21, 2024. https://www.rhythmresearchresources.net/research-drum-therapy-introduction.html.

Schlaug, Gottfried. "Musicians and Music Making as a Model for the Study of Brain Plasticity." *Progress in Brain Research* 217 (2015): 37–55. https://doi.org/10.1016/bs.pbr.2014.11.020.

Schlaug, Gottfried. "The Brain of Musicians." In *The Cognitive Neuroscience of Music*, edited by Isabelle Peretz and Robert Zatorre, 366–81. Oxford University Press, 2003.

Schlaug, Gottfried, Andrea Norton, Sarah Marchina, Lauryn Zipse, and Catherine Wan. "From Singing to Speaking: Facilitating Recovery from Nonfluent Aphasia." *Future Neurology* 5, no. 5 (2010): 657–65. https://doi.org/10.2217/fnl.10.44.

Schlund, Michael, and Michael Cataldo. "Amygdala Involvement in Human Avoidance, Escape and Approach Behavior." *Neuroimage* 53, no. 2 (2010): 769–76.https://doi.org/10.1016/j.neuroimage.2010.06.058.

Schmitz, Gerd, Jeannine Bergmann, Alfred Effenberg, Carmen Krewer, Tong-Hun Wang, and Friedemann Müller. "Movement Sonification in Stroke Rehabilitation." *Frontiers in Neurology* 9 (2018): article 389. https://doi.org/10.3389/fneur.2018.00389.

Scholz, Daniel, Sönke Rohde, Nikou Nikmaram, Hans-Peter Brückner, Michael Großbach, Jens D. Rollnik, and Eckart Altenmüller. "Sonification of Arm Movements in Stroke Rehabilitation—A Novel Approach in Neurologic Music Therapy." *Frontiers in Neurology* 7 (2016): article 106. https://doi.org/10.3389/fneur.2016.00106.

Schubert, Leda. *Listen: How Pete Seeger Got America Singing*. Roaring Brook Press, 2017.

Seeley, William, Richard Crawford, Juan Zhou, Bruce Miller, and Michael Greicius. "Neurodegenerative Diseases Target Large-Scale Human Brain

Networks." *Neuron* 62, no. 1 (2009): 42–52. https://doi.org/10.1016/j
.neuron.2009.03.024.

Segelman, Mikhail. "Vissarion Shebalin." *Wise Music Classical*, n.d. https://
www.wisemusicclassical.com/composer/2984/Vissarion-Shebalin/.

Service, Tom. "The Ice Age Flute That Can Play the Star-Spangled Banner."
Guardian, February 15, 2013. https://www.theguardian.com/music/2013
/feb/15/ice-age-flute.

Sihvonen, Aleksi, Pablo Ripollés, Vera Leo, Antoni Rodríguez-Fornells, Seppo
Soinila, and Teppo Särkämö. "Neural Basis of Acquired Amusia and Its
Recovery after Stroke." *Journal of Neuroscience* 36, no. 34 (2016): 8872–
81. https://doi.org/10.1523/JNEUROSCI.0709-16.2016.

Sihvonen, Aleksi, Teppo Särkämö, Antoni Rodríguez-Fornells, Pablo Ripollés,
Thomas Münte, and Seppo Soinila. "Neural Architectures of Music—
Insights from Acquired Amusia." *Neuroscience & Biobehavioral Reviews* 107
(2019): 104–14. https://doi.org/10.1016/j.neubiorev.2019.08.023.

Sihvonen, Aleksi, Teppo Särkämö, Vera Leo, Mari Tervaniemi, Eckart Altenmüller, and Seppo Soinila. "Music-Based Interventions in Neurological
Rehabilitation." *Lancet Neurology* 16, no. 8 (2017): 648–60. https://doi.org
/10.1016/S1474-4422(17)30168-0.

Smee, Sebastian. "MFA Expands Loans of Well-Known Works." *Boston Globe*,
August 2, 2014. https://www.bostonglobe.com/arts/2014/08/02/museum
-fine-arts-controversial-lending-program-keeps-works-off-walls
/9dsHEU1Faxh8AbKbGRN6LL/story.html.

Soley, Gaye, and Erin Hannon. "Infants Prefer the Musical Meter of Their
Own Culture: A Cross-Cultural Comparison." *Developmental Psychology*
46, no. 1 (2010): 286–92. https://doi.org/10.1037/a0017555.

Somayaji, Kamila, Mogen Frenkel, Luai Tabaza, Alexis Visotcky, Tanya Ruck,
Ernest Ofori, Michael Widlansky, and Jacquelyn Kulinski (2022). "Acute
Effects of Singing on Cardiovascular Biomarkers." *Frontiers in Cardiovas-
cular Medicine* 9 (2022): 869104. https://doi.org/10.3389/fcvm.2022.869104.

Spitzer, Michael. *The Musical Human: A History of Life on Earth*. Bloomsbury,
2021.

startv9953. "Baby React [sic] to Music." YouTube October 18, 2022. https://
www.youtube.com/shorts/SJ9OnIxtEQE.

Stavropoulos, Jenn. "Baby Gets Emotional When Mom Sings Opera!" You-
Tube, July 27, 2018. 2018). https://www.youtube.com/watch?v=jME9Q
yhHS_0.

Stewart, Lauren, Katharina von Kriegstein, Jason Warren, and Timothy Griffiths. "Music and the Brain: Disorders of Musical Listening." *Brain* 129, no. 10 (2006): 2533–53. https://doi.org/10.1093/brain/awl171.

Sulzer, David. *Music, Math, and Mind: The Physics and Neuroscience of Music.* Columbia University Press, 2021.

Svard, Lois. "Music and Prehistoric Cave Art," June 28, 2018. https://www.themusiciansbrain.com/music-and-prehistoric-cave-art/.

Tanakh: A New Translation of the Holy Scriptures According to the Traditional Hebrew Text. Jewish Publication Society, 1985.

"10 Lessons Your Child Will Learn in Marching Band." *Music & Arts*, February 28, 2022. https://thevault.musicarts.com/10-lessons-your-child-will-learn-in-marching-band/.

Thaut, Michael. "Coda and Crescendo: How Neuroscience Created Neurologic Music Therapy to Help Heal the Injured Brain." In Fleming, *Music and Mind.*

Thaut, Michael. "The Discovery of Human Auditory-Motor Entrainment and Its Role in the Development of Neurologic Music Therapy." *Progress in Brain Research* 217 (2015): 253–66. https://doi.org/10.1016/bs.pbr.2014.11.030.

Todd, Neil, and Christopher Lee. "The Sensory-Motor Theory of Rhythm and Beat Induction 20 Years On: A New Synthesis and Future Perspectives." *Frontiers in Human Neuroscience* 9 (2015): article 444. https://doi.org/10.3389/fnhum.2015.00444.

Tomaino, Concetta. *Music Has Power® in Senior Wellness and Healthcare: Best Practices from Music Therapy.* Jessica Kingsley, 2023.

Trainor, Laurel, Xiaoqing Gao, Jing-jiang Lei, Karen Lehtovaara, and Laurence Harris. "The Primal Role of the Vestibular System in Determining Musical Rhythm." *Cortex* 45, no. (2009): 35–43.

Trehub, Sandra. "The Developmental Origins of Musicality." *Nature Neuroscience* 6, no. 7 (2003):669–73.

Trehub, Sandra. "Infant Musicality." In *The Oxford Handbook of Music Psychology,* 2nd ed., edited by Susan Hallam, Ian Cross, and Michael Thaut, 387–97. Oxford University Press, 2016.

Trehub, Sandra. "Musical Predispositions in Infancy: An Update." In *The Cognitive Neuroscience of Music,* edited by Isabelle Peretz and Robert Zatorre, 3–20. Oxford University Press, 2003.

Trehub, Sandra. "Musical Universals: Perspectives from Infancy." In *Actualité des Universaux musicaux (Topics in Musical Universals)*, edited by J. L. Leroy, 5–8. Éditions des archives contemporaines, 2013.

Trehub, Sandra, Judith Becker, and Iain Morley. "Cross-Cultural Perspectives on Music and Musicality." *Philosophical Transactions of the Royal Society B* 370 (2015): 20140096. https://doi.org/10.1098/rstb.2014.0096.

University of Liverpool, Acoustics Research Unit. "Musical Vibrations: Using Vibrotactile Technology to support d/Deaf People in Music Education at the Royal School for the Deaf Derby." *liverpool.cloud.panopto.eu*. (n.d.). https://stream.liv.ac.uk/2qvwd9th.

Van der Meulen, Ineke, Mieke van de Sandt-Koenderman, and Gerard Ribbers. "Melodic Intonation Therapy: Present Controversies and Future Opportunities." *Archives of Physical Medicine and Rehabilitation* 93, no. 1 (supp. 1) (2012): S46–S52. https://doi.org/10.1016/j.apmr.2011.05.029.

Vander Elst, Olivia, Nicholas Foster, Peter Vuust, and Morten Kringelbach. "The Neuroscience of Dance: A Systematic Review of the Present State of Research and Suggestions for Future Work" *Neuroscience and Behavioral Reviews*, preprint: 2021. https://doi.org/10.1016/j.neubiorev.2023.105197.

Vassena, Elianna, Clay Holroyd, and William Alexander. "Computational Models of Anterior Cingulate Cortex: At the Crossroads Between Prediction and Effort." *Frontiers in Neuroscience* 11 (2017): article 316. https://doi.org/10.3389/fnins.2017.00316.

Verghese, Joe, Aaron LeValley, Carol Derby, Gail Kuslansky, Mindy Katz, Charles Hall, Herman Buschke, and Richard Lipton. "Leisure Activities and the Risk of Amnestic Mild Cognitive Impairment in the Elderly." *Neurology* 66, no. 6 (2006): 821–27. https://doi.org/10.1212/01.wnl.0000202520.68987.48.

Verghese, Joe, Richard Lipton, Mindy Katz, Charles Hall, Carol Derby, Gail Kuslansky, Anne Ambrose, Martin Sliwinski, and Herman Buschke. "Leisure Activities and the Risk of Dementia in the Elderly." *New England Journal of Medicine* 348, no. 25 (2003): 2508–16. https://doi.org/10.1056/NEJMoa0222.

Victor Café. "History: Our Story." https://victorcafe.com/History-2.

Viskontas, Indre. *Child*. San Francisco Conservatory of Music and the Anne and Gordon Getty Foundation, 2020. https://sfcm.edu/sites/default/files/SFCM-Music_for_Every_Child.pdf.

Vlaicu, Andrei, and Mihaela Bustuchina. "La musique, un bon outil pour étudier le cerveau." *PSN* 17, no. 2 (2019): 27–37. https://doi.org/10.3917/psn.172.0027.

Wan, Catherine, Theodor Rüber, Anja Hohmann, and Gottfired Schlaug. "The Therapeutic Effects of Singing in Neurological Disorders." *Music Perception* 27, no. 4 (2010): 287–95. https://doi.org/10.1525/mp.2010.27.4.287.

Washington, Stuart, and John Tillinghast. "Conjugating Time and Frequency: Hemispheric Specialization, Acoustic Uncertainty, and the Mustached Bat." *Frontiers in Neuroscience* 9 (2015): article 143. https://doi.org/10.3389/fnins.2015.00143.

Watson, Christopher. "The Songs of Gods and Men: Internal Songs and Singers in Archaic Greek Epic." Master's thesis, University of Kansas, 2016. https://hdl.handle.net/1808/21910.

Watson, Sarah. "Critical Review: Melodic Intonation Therapy. The Influence of Pitch and Rhythm on Therapy Outcomes." University of Western Ontario, 2016. https://www.uwo.ca/fhs/lwm/teaching/EBP/2015_16/Watson.pdf.

Whaling, Carol. "What's Behind a Song? The Neural Basis of Song Learning in Birds." In *The Origins of Music,* edited by Nils Wallin, Björn Merker, and Steven Brown, 65–76. MIT Press, 2000.

Winsler, Adam, Lesley Ducenne, and Abel Koury. "Singing One's Way to Self-Regulation: The Role of Early Music and Movement Curricula and Private Speech." *Early Education and Development* 22 (2011): 274–304. http://dx.doi.org/10.1080/10409280903585739.

Witynski, Max. "Perfect Pitch, Explained." *uchicago.edu.* (Date not indicated): https://news.uchicago.edu/explainer/what-is-perfect-pitch#:~:text=A%20commonly%20cited%20number%20is,thought%20to%20have%20perfect%20pitch.

Wurz, Sarah. "Interpreting the Fossil Evidence for the Evolutionary Origins of Music." *Southern African Humanities* 21 (2009): 395–417. https://hdl.handle.net/10520/EJC84839.

Yun, Paul. "Music, Rhythm, and the Brain." *BrainWorld,* January 22, 2022. https://brainworldmagazine.com/music-rhythm-brain/.

Zatorre, Robert. *From Perception to Pleasure: The Neuroscience of Music and Why We Love It.* Oxford University Press, 2024.

Zatorre, Robert. "Music, the Food of Neuroscience?" *Nature* 434 (2005): 312–15. https://doi.org/10.1038/434312a.

Zatorre, Robert. "Musical Enjoyment and the Reward Circuits of the Brain," In *Music and Mind: Harnessing the Arts for Health and Wellness*, edited by Renée Fleming, 434–49. Viking, 2024.

Zatorre, Robert. "Neural Specializations for Tonal Processing." In *The Cognitive Neuroscience of Music*, edited by Isabelle Peretz and Robert Zatorre, 231–46. Oxford University Press, 2003.

Zatorre, Robert. "Why Do We Love Music?" *Cerebrum*, November 1, 2018, cer-1618. https://pmc.ncbi.nlm.nih.gov/articles/PMC6353111/pdf/cer-16-18.pdf.

Zatorre, Robert, Pascal Belin, and Virginia Penhune. "Structure and Function of Auditory Cortex: Music and Speech." *Trends in Cognitive Sciences* 6 (2002): 37–46. https://doi.org/10.1016/S1364-6613(00)01816-7.

Zatorre, Robert, Joyce Chen, and Virginia Penhune. "When the Brain Plays Music: Auditory-Motor Interactions in Music Perception and Production." *Nature Reviews/Neuroscience* 8, no. 7 (2007): 547–58. https://doi.org/10.1038/nrn2152.

Zentner, Marcel, and Tuomas Eerola. "Rhythmic Engagement with Music in Infancy." *Proceedings of the National Academy of Sciences* 107, no. 13 (2010): 5768–73. https://doi.org/10.1073/pnas.1000121107.

Zhou, Wenjie, Chonghuan Ye, Haitao Wang, Yu Mao, Weijia Zhang, An Liu, Chen-Ling Yang, Tianming Li, Lauren Hayashi, Wan Zhao, Lin Chen, Yuanyuan Liu, Wenjuan Tao, and Zhi Zhang. "Sound Induces Analgesia Through Corticothalamic Circuits." *Science* 377, no. 6602) (2022): 198–204. https://doi.org/10.1126/science.abn466.

Index

About the Author

Dr. Samuel Markind has long been interested in the brain. So, it was a natural choice for him to specialize in neurology after he finished medical school. During his more than 30 years in practice, he had a particular interest in evaluating and treating patients with dementia. This included participating in clinical trials of medications being developed for memory disorders. Though now retired, he continues to serve on the Medical Scientific Advisory Committee of the Connecticut Chapter of the Alzheimer's Association.

The combination of neurology and love of music fuels Dr. Markind's desire to share his understanding of the relationship between music and the brain with others. This passion led him to develop continuing medical education presentations that musically explore the brain basis of creativity. It also inspired him to write this book focusing on the evolutionary impact of music upon the human brain. His goals are to illuminate ways that music benefits the brain as well as to encourage people to engage actively with music.

Sam and his wife, Dina, have three grown children and became grandparents in 2022. They live in Connecticut and enjoy dancing together.

Follow/connect with Dr. Markind on LinkedIn at https://www .linkedin.com/in/sam-markind-b67718136 or through his website: www.musicbetweenyourears.com

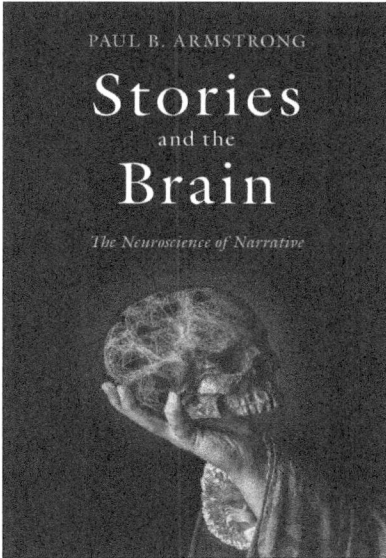

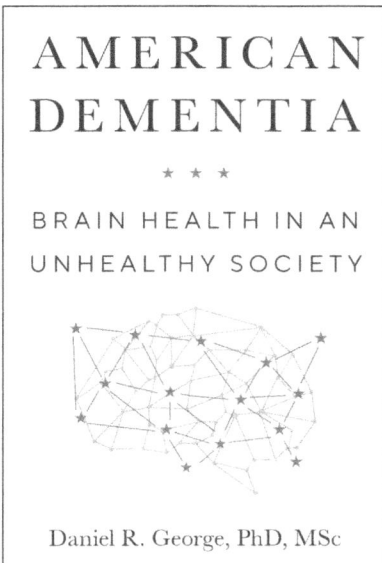

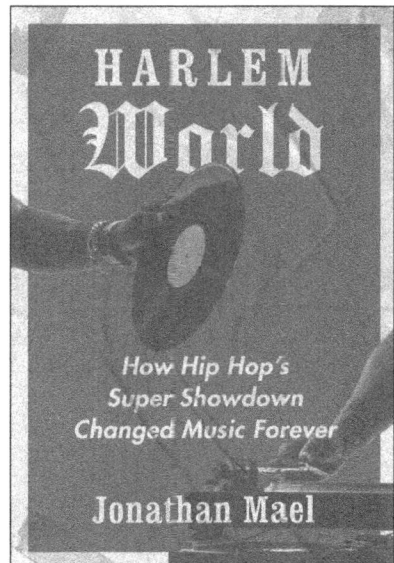

.